Activation and Detoxification Enzymes

Chang-Hwei Chen

Activation and Detoxification Enzymes

Functions and Implications

Second Edition

Chang-Hwei Chen
Institute for Health, and the Environment and Department of Biomedical Sciences
University at Albany, State University of New York
Albany, NY, USA

ISBN 978-3-031-55286-1   ISBN 978-3-031-55287-8 (eBook)
https://doi.org/10.1007/978-3-031-55287-8

© The Editor(s) (if applicable) and The Author(s), under exclusive license to Springer Nature Switzerland AG 2012, 2024

This work is subject to copyright. All rights are solely and exclusively licensed by the Publisher, whether the whole or part of the material is concerned, specifically the rights of translation, reprinting, reuse of illustrations, recitation, broadcasting, reproduction on microfilms or in any other physical way, and transmission or information storage and retrieval, electronic adaptation, computer software, or by similar or dissimilar methodology now known or hereafter developed.

The use of general descriptive names, registered names, trademarks, service marks, etc. in this publication does not imply, even in the absence of a specific statement, that such names are exempt from the relevant protective laws and regulations and therefore free for general use.

The publisher, the authors, and the editors are safe to assume that the advice and information in this book are believed to be true and accurate at the date of publication. Neither the publisher nor the authors or the editors give a warranty, expressed or implied, with respect to the material contained herein or for any errors or omissions that may have been made. The publisher remains neutral with regard to jurisdictional claims in published maps and institutional affiliations.

This Springer imprint is published by the registered company Springer Nature Switzerland AG
The registered company address is: Gewerbestrasse 11, 6330 Cham, Switzerland

Paper in this product is recyclable.

*This book is dedicated to
my father and my mother*

Preface

Many advances have been made in the past decades toward the understanding of the functions and implications of activation and detoxification enzymes. The present edition represents the efforts to update their general overviews, including (a) foreign compounds that humans are exposed, (b) metabolisms catalyzed by activation and detoxification enzymes, (c) reactive chemical intermediate generated from foreign compound metabolisms, (d) oxidative stress derived from reactive chemical intermediates, (e) oxidative stress-induced cell toxicities and health effects, (f) inducibility of metabolic enzymes for health benefits, and (g) effects of dietary inducers on metabolic enzymes.

Humans are exposed to foreign compounds, such as foods, drugs, household products, and environmental chemicals. Among these foreign compounds of significance is that they mainly belong to non-water-soluble substances. Metabolic conversion of such lipophilic compounds to water-soluble species in order to facilitate their removal from the body is carried out by biochemical reactions, where functionalization reaction catalyzed by activation enzyme introduces a functional group to increase the aqueous solubility of a lipophilic foreign compound. Functionalized compounds are then further metabolized through detoxification enzyme-catalyzed reactions to inactivate metabolic intermediates and to facilitate their excretion from the body.

Importantly, reactive chemical intermediates formed during metabolic activation reactions are capable of interacting with cellular components to cause damages. An imbalance between the production of reactive chemical intermediates and the ability of organisms to detoxify them can cause oxidative stress, which can lead to various disease conditions. The expressions of metabolic enzymes exhibit genetic polymorphisms among individuals. Lifestyle factors can also affect an individual's susceptibility to xenobiotics-mediated toxic effects.

In addition, a number of chemical compounds are capable of acting as modulators for activation and detoxification enzymes. Many of these chemical compounds are present in human daily diets. Modulation of metabolic enzymes as a useful approach for human health benefits is an active research area.

This book addresses the above subjects in an updated and a broad scope manner with emphasis on biochemical and biomedical aspects. The book can be used as text materials for senior undergraduates and graduate students in the areas of biochemistry, biomedical sciences, pharmacology, and environmental sciences. Researchers may also be interested in this book to gain knowledge on the advances in foreign compound metabolisms and their implications.

The author wishes to thank Professor David Carpenter, Director of the Institute of Health and the Environment, the University at Albany, for his many courtesies and suggestions, and Professor Norman L. Strominger of Albany Medical College and the University at Albany for his valuable comments and inputs. The author also wishes to express appreciation to the editorial board at Springer Nature for their assistance.

Albany, NY, USA   Chang-Hwei Chen

Introduction

Humans digest foreign compounds, including food, household products, pharmaceuticals, environmental chemicals, and lifestyle products, through enzyme catalytic reactions. During activation metabolism to enhance aqueous solubility of foreign compound, a significant number of reactive chemical intermediates are generated, which has the potential to cause oxidative stress, resulting in various disease conditions. Detoxification metabolism catalyzed by detoxification enzyme is then followed to detoxify foreign compounds and speedily remove them from the body.

Many advances have been made toward the understanding of foreign compound metabolisms during the last decades, including various foreign compounds that humans are exposed, metabolic activation and detoxification enzymes, reactive chemical intermediates or metabolites, inducibility of metabolic enzymes, dietary inducers of metabolic enzymes, and inducer-drug interactions.

Metabolic enzymes catalyze a broad spectrum of reactions, including oxidation, hydrolysis, reduction, and conjugation and non-conjugation reactions. Understanding these catalytic reactions is essential for investigating reactive intermediates generated during foreign compound metabolism. Reactive intermediates are either electrophiles or nucleophiles. Electrophile prefers to interact with a nucleophilic group, while a nucleophile prefers to interact with an electrophilic group. Their catalytic reactions represent critical elements of foreign compound metabolic processes.

Oxidative stress is characterized by an imbalance between the production of reactive chemical intermediates and the ability of organisms to neutralize them. Knowledge about reactive intermediates and their underlying mechanisms are critical to the understanding of foreign compound-mediated toxic effects. Inflammation is a biological response to oxidative stress. Oxidative stress and chronic inflammation play a significant role in underlying many disease conditions and initiation of carcinogenesis.

Detoxification metabolisms that function correctly appear to be a vital means of preventing foreign compound-mediated toxic effects. To minimize such exposure, it is essential to maintain reactive intermediates or metabolites at minimum levels. This maintenance requires a delicate balance between activation and detoxification enzyme systems. A fine balance is dependent on relative efficiencies of these enzymes. Accordingly, living cells maintain an inducible antioxidant response pathway regulated by nuclear factor erythroid-2-related factor 2.

Two important hypotheses have been proposed to maintain reactive intermediates or metabolites at minimum levels. One hypothesis is to inhibit activation enzymes along with inducing detoxification enzymes. Another hypothesis proposes that the induction of detoxification enzymes alone is enough to provide a measure against reactive chemical intermediates generated from foreign compound metabolism. Generally, an unusual high expression of activation enzymes may give rise to an overload of reactive intermediates or metabolites, while an extraordinary low efficacy of detoxification enzyme may result in abnormal low efficiency in detoxification of foreign compounds.

Modulation of metabolic enzymes as a useful approach for reducing foreign compound-mediated toxic effects has been a subject of intense interest. Advances in the understanding of mechanisms that govern the detoxification of foreign compounds that human digest have revealed that diets can have significant impacts on the efficacies of metabolic enzymes.

Extensive investigations in the past decades have discovered a variety of chemical compounds that are capable of acting as inducers or inhibitors for metabolic enzymes. Lists of such modulation compounds are included in this book. Many chemical compounds in human diets are capable of acting as metabolic enzyme modulators.

Such inducers that are present in vegetables, fruits, herbs, beverages, and algae have received much attention. In particular, those that are: rich in detoxification enzymes, such as uridine-diphosphate-glucuronosyl-transferase and glutathione S-transferase. There are substantial evidences to support the hypothesis that intaking diet rich in detoxification enzyme inducers is a promising proposal to minimize foreign compound-mediated toxic effects.

Extensive lists of vegetables and fruits that are rich in detoxification inducers can be found in this book. Such inducers of detoxification enzymes consist of a variety of chemical classes. Structural investigations of enzyme modulators reveal that many inducers of detoxification enzymes contain Michael acceptor functionalities, such as olefins or acetylenes. These findings may provide important insights into future developments of new enzyme inducers. A modulator molecule may affect the activity of an enzyme by changing its conformation or by interfering with substrate-enzyme interactions. Research on potential enzyme conformation change caused by inducer interaction requires future attention.

Moreover, the expression of metabolic enzymes may vary among individuals. Genetic polymorphisms are an important factor in contributing to individual variations in the efficacies of metabolic enzymes, particularly, activation enzyme such as cytochrome P450s and detoxification enzyme such as glutathione-S-transferases. Variations in genetic polymorphisms can affect an individual's susceptibility to foreign compound-mediated toxic effects. Environmental and lifestyle factors, including cigarette smoking and alcohol drinking, can also affect the expressions of metabolic enzymes. These important issues are also discussed in the book.

Contents

1 **Overview** 1
   1.1 Foreign Compounds That Humans Expose 2
   1.2 Metabolic Reactions of Foreign Compounds 2
   1.3 Activation Enzymes 2
   1.4 Detoxification Enzymes 2
   1.5 Reactive Metabolic intermediates 3
   1.6 Oxidative Stress, Electrophiles, and Free Radicals 3
   1.7 Lifestyle Modifications 3
   1.8 Metabolic Intermediate-Induced Cell Toxicities 4
   1.9 Genetic Polymorphism of Metabolic Enzymes 4
   1.10 Defenses Against Metabolic Intermediates 4
      1.10.1 Inducibility of Metabolic Enzymes 4
      1.10.2 Diversified Classes of Enzyme Modulators 5
      1.10.3 Induction and Inhibition Compound 5
   1.11 Defense Mechanisms: Nrf2-ARE Pathway 5
   1.12 Health Effects of Metabolic Intermediates 5
   1.13 Induction of Metabolic Enzymes for Health Benefits 6
   1.14 Dietary Effects on Metabolic Enzymes 6
   Bibliography 7

2 **Foreign Compounds: Foods, Drugs, Chemicals, and Lifestyles** 9
   2.1 Foods 9
      2.1.1 Heterocyclic Amines 10
      2.1.2 Nitrosamines 10
      2.1.3 Polycyclic Aromatic Hydrocarbons 11
      2.1.4 Azo Dyes 11
      2.1.5 α,β-Unsaturated Aldehydes 12
      2.1.6 Mycotoxin 12
   2.2 Household Products 12
      2.2.1 Benzene 13
      2.2.2 Phenol 13
      2.2.3 Phthalate 13

	2.3	Pharmaceuticals	13
		2.3.1 Acetaminophen	14
		2.3.2 Xanthine	14
		2.3.3 Terfenadine	14
		2.3.4 Menadione	15
		2.3.5 Diazepam	15
	2.4	Environmental Chemicals	15
		2.4.1 Diesel Exhausts	15
		2.4.2 Arsenic	16
		2.4.3 Polychlorinated Biphenyls	16
		2.4.4 Dioxins	16
	2.5	Lifestyles	17
		2.5.1 Alcohol	17
		2.5.2 Cigarette	17
		Bibliography	18
3	**Transport and Excretion of Foreign Compounds**		**21**
	3.1	Lipophiles Versus Hydrophiles	22
	3.2	Hydrogen Bonding	23
	3.3	Sites of Action	23
	3.4	Cell Membranes	24
	3.5	Transport Mechanisms	24
		3.5.1 Passive Diffusion	25
		3.5.2 Facilitated Diffusion	25
		3.5.3 Active Transport	26
	3.6	Metabolic Pathways	26
		3.6.1 Phase I Activation Metabolism	27
		3.6.2 Phase II Detoxification Metabolism	27
	3.7	Transport to External Cell Compartment	28
	3.8	Metabolism Precedes Excretion	28
	3.9	Excretion of Foreign Compounds	29
		3.9.1 Renal Excretion	29
		3.9.2 Reabsorption in the Kidney	30
		3.9.3 Hepatic Excretion	30
		3.9.4 Skin Excretion	31
		Bibliography	31
4	**Metabolic Conversion of Foreign Compounds**		**33**
	4.1	Phase I Activation Metabolism	34
		4.1.1 Oxidation Reactions	34
		4.1.2 Hydroxylation Reactions	34
		4.1.3 Dealkylation Reactions	35
		4.1.4 Hydrolysis	35
		4.1.5 Epoxidation	36
	4.2	Phase II Detoxification Metabolism	36
		4.2.1 Conjugation Reactions	37
		4.2.2 Non-conjugation Reactions	39

4.3	Toxification and Detoxification	40
	4.3.1 Toxification Activation	40
	4.3.2 Deactivation of Toxicity	40
	4.3.3 Activation Versus Deactivation: Competing Pathways	41
Bibliography		42

5 Phase I Activation Enzymes ... 45
5.1 Activation of Foreign Compounds ... 46
5.2 Activation Enzymes ... 46
5.2.1 Oxidative Enzymes ... 47
5.2.2 Reductive Enzymes ... 51
5.2.3 Hydrolytic Enzymes ... 51
5.3 Catalytic Reactions ... 52
5.3.1 Oxidative Reactions ... 52
5.3.2 Reductive Reactions ... 54
5.3.3 Hydrolytic Reaction ... 55
Bibliography ... 56

6 Phase II Detoxification Enzymes ... 59
6.1 Exclusion of Foreign Compounds ... 59
6.2 Detoxification Enzymes ... 60
6.3 Conjugation Enzymes ... 60
6.3.1 Uridine-Diphosphate-Glucuronosyltransferase ... 60
6.3.2 Glutathione S-Transferase ... 61
6.3.3 Sulfotransferase ... 62
6.3.4 N-Acetyltransferase ... 63
6.3.5 Methyltransferase ... 63
6.3.6 Acyltransferase ... 64
6.4 Conjugation Enzyme-Catalyzed Reactions ... 64
6.4.1 Conjugation at O Atom ... 64
6.4.2 Conjugation at N Atom ... 65
6.4.3 Conjugation at C Atom ... 65
6.4.4 Conjugation at S Atom ... 66
6.4.5 Conjugation of Carboxylic Acid ... 66
6.4.6 Conjugation at OH Group ... 67
6.5 Non-conjugation Enzymes ... 67
6.5.1 Quinone Reductase ... 68
6.5.2 Epoxide Hydrolases ... 68
6.6 Non-conjugation Enzyme-Catalyzed Reactions ... 68
6.6.1 Quinone Reductase ... 68
6.6.2 Epoxide Hydrolase ... 69
Bibliography ... 69

7 Catalytic Reactions of Activation Enzymes ... 71
7.1 Cytochrome P450 ... 71
7.1.1 Hydroxylation of Aliphatic or Aromatic Compounds ... 72

		7.1.2	Epoxidation of Ether	72
		7.1.3	Dehydrogenation of Alcohol or Aldehyde	73
		7.1.4	Oxidation of N - or S - Compound	73
		7.1.5	Dealkylation of Ether, Amide, or Carboxylic Acid	74
		7.1.6	Oxidation of Carbon on Aromatic Ring	74
		7.1.7	Activation of Benzo[a]pyrene	74
	7.2	Prostaglandin H Synthase		74
	7.3	Flavin Monooxygenase		75
	7.4	Amine Oxidase		75
	7.5	Nitroreductase		76
	7.6	Azoreductase		76
	7.7	Molybdenum Hydroxylase		76
	7.8	Alcohol Dehydrogenase		77
	7.9	Ribonucleotide Reductase		77
	7.10	Peroxidase		77
	7.11	Carboxylesterase		78
	Bibliography			78
8	**Catalytic Reactions of Detoxification Enzymes**			81
	8.1	Conjugation Reactions		82
		8.1.1	Uridine Diphospho-Glucuronosyl Transferase	82
		8.1.2	Glutathione S-Transferase	83
		8.1.3	Sulfotransferase	84
		8.1.4	Acyltransferase	85
		8.1.5	N-Acetyltransferase	85
		8.1.6	Methyltransferase	86
	8.2	Non-conjugation Reactions		87
		8.2.1	Quinone Reductase	87
		8.2.2	Epoxide Hydrolase	87
	8.3	Catalytic Actions on Atom or Molecule		88
		8.3.1	Conjugation at O Atom	88
		8.3.2	Conjugation at N Atom	89
		8.3.3	Conjugation at C Atom	90
		8.3.4	Conjugation at S Atom	90
		8.3.5	Conjugation of Carboxylic Acid	91
	8.4	Other Non-conjugation Reactions		91
	Bibliography			92
9	**Reactive Intermediates and Their Interactions**			95
	9.1	Reactive Intermediate Species		95
		9.1.1	Reactive Oxygen Species	97
		9.1.2	Reactive Nitrogen Species	97
	9.2	Enzyme-Catalyzed Reactive Intermediate Formation		98
		9.2.1	Mediation by Activation Enzymes	98
		9.2.2	Mediation by Detoxification Enzymes	99
	9.3	Interactions with Cellular Components		100
		9.3.1	Protein Adducts	101

		9.3.2	DNA Adducts	101
		9.3.3	Lipid Peroxidation	101
	9.4	Factors Affecting Foreign Compound Toxicity		102
	9.5	Defense Against Reactive Intermediates		102
		9.5.1	Conjugation Reactions	102
		9.5.2	Glutathione	103
		9.5.3	Antioxidant Enzymes	103
	Bibliography			104
10	**Metabolite-Associated Cell Toxicities**			107
	10.1	Intrinsic Toxicity		107
	10.2	Reactive Intermediate–Related Toxicity		108
	10.3	Lifestyle-Induced Toxicity		108
		10.3.1	Alcohol	109
		10.3.2	Cigarette	109
	10.4	Toxic Effects on Cell Components		109
		10.4.1	Protein Damage	110
		10.4.2	DNA Damage	111
		10.4.3	Lipid Peroxidation	111
	10.5	Toxic Effects on Cellular Functions		112
		10.5.1	Intervention with Mitochondria Functions	112
		10.5.2	Interaction with Ion Transporters	113
		10.5.3	Interference with Enzymatic Functions	113
		10.5.4	Immune Suppression and Stimulation Effects	114
	10.6	Chemical Carcinogenesis		115
	10.7	Drug Metabolism Interference		115
	Bibliography			116
11	**Oxidative Stress: Reactive Chemical Intermediates**			119
	11.1	Oxidative Stress		120
		11.1.1	Reactive Oxygen Species	120
		11.1.2	Free Radicals	121
	11.2	Reactive Intermediate–Mediated Oxidative Stress		121
		11.2.1	Oxidative Stress on Biomolecules	121
		11.2.2	Oxidative Stress and Inflammation on Diseases	123
	11.3	Electrophilic Stress		124
		11.3.1	Foreign Compound–Mediated Electrophiles	124
		11.3.2	Varieties of Electrophiles	125
		11.3.3	Electrophilic Stress on Biomolecules	125
		11.3.4	Electrophiles on Disease Conditions	126
		11.3.5	Electrophiles on Drug Metabolism	127
	11.4	Defense Against Oxidative Stress		127
		11.4.1	Chemoprevention Inducers	127
		11.4.2	Nrf2- ARE Pathway	128
	Bibliography			128

12 Metabolic Enzymes: Polymorphisms and Species Differences 131
12.1 Enzyme Polymorphisms on Xenobiotic Metabolism 132
12.2 Genetic Polymorphisms of Activation Enzymes 132
    12.2.1 Cytochrome P450 (CYP450) Polymorphisms 133
    12.2.2 Flavin-Containing Monooxygenase 136
    12.2.3 Peroxidase 137
    12.2.4 Carboxylesterase 137
    12.2.5 Alcohol and Aldehyde Dehydrogenases Polymorphisms 137
12.3 Genetic Polymorphisms of Detoxification Enzymes 138
    12.3.1 Glutathione S Transferase (GST) 139
    12.3.2 UDP-Glucuronosyltransferase 140
    12.3.3 Sulfotrasferase 140
    12.3.4 N-Acetyltransferase 141
    12.3.5 Methyltransferase 141
    12.3.6 Quinone Oxidoreductase 142
    12.3.7 Epoxide Hydrolase 143
12.4 Enzyme Polymorphisms on Alcohol and Smoke 143
    12.4.1 Alcoholism 143
    12.4.2 Smoker 144
12.5 Species Difference in Metabolic Enzyme Activities 144
    12.5.1 Susceptibility to Aflatoxin Toxicity between Humans and Mice 145
    12.5.2 Resistance to Tamoxifen Toxicity in Humans, Not in Rats 145
    12.5.3 Different 4-Ipomeanol Toxicity Between Humans and Rodents 146
Bibliography 147

13 Defense Against Oxidative Stress: Nrf2-ARE Pathway 151
13.1 Transcription Factor Nrf2 151
    13.1.1 Role of Nrf2 on Oxidative Stress 152
    13.1.2 Keap1 Regulation of Nrf2 Activity 153
13.2 Activation of Nrf2-ARE Pathway 153
13.3 Nrf2-KeapP1-ARE Pathway 153
13.4 Molecular Mechanism of Nrf2-ARE Pathway 154
    13.4.1 In the Absence of Oxidative Stress 154
    13.4.2 In the Presence of Oxidative Stress 155
13.5 Cytoprotection Through Nrf2-ARE Pathway 155
    13.5.1 Induction of Nrf2 – ARE Pathway 156
    13.5.2 Enzyme Inducers for Cytoprotection 156
    13.5.3 Over-Activation of Nrf2-ARE Pathway 157
13.6 Role of Nrf2 in Diseases 158
13.7 Nrf2-Inducing Compounds for Chemoprevention 158
Bibliography 159

14	**Inducibility of Metabolizing Enzymes**		161
	14.1	Inducibility of Activation Enzymes	162
	14.2	Inducibility of Detoxification Enzymes	162
	14.3	Lifestyle Modification	164
	14.4	Monofunctional and Bifunctional Inducers	164
	14.5	Balance Between Activation and Detoxification Metabolisms	165
	14.6	Enzyme Modulation Against Potential Toxicity	166
		14.6.1 Enzyme Modulation	167
		14.6.2 Hypothesis of Detoxification Enzyme Induction	167
	14.7	Inducer–Metabolic Enzyme Interactions	167
		14.7.1 Michael Reaction Acceptors	168
		14.7.2 Unsaturated Carbon–Carbon Bonds	169
		14.7.3 Phenolic Hydroxyl Groups	170
	Bibliography		171
15	**Inducibility of Metabolizing Enzymes**		173
	15.1	Defense Against Potential Metabolic Toxicity	174
		15.1.1 Modification of Activation Enzymes	174
		15.1.2 Modification of Detoxification Enzymes	174
		15.1.3 Antioxidant Activities	175
	15.2	Modification of Metabolic Enzymes	175
		15.2.1 Modification of Activation Enzymes	175
		15.2.2 Modulation of Detoxification Enzymes	176
	15.3	Major Inducers of Metabolic Enzymes	177
		15.3.1 Sulforaphane and Other Isothiocyanates	178
		15.3.2 1,2-Dithiole-3-Thione and Derivatives	178
		15.3.3 Indole-3-Carbinol	180
		15.3.4 Flavonoids and Isoflavones	180
		15.3.5 Polyphenols	182
		15.3.6 Organosulfur Compounds	183
		15.3.7 Terpenes and Terpenoids	184
	15.4	Other Inducers	185
		15.4.1 Phenobarbital	185
		15.4.2 Grapefruit	185
	Bibliography		186
16	**Diversified Classes of Enzyme Modulators**		189
	16.1	Substrate–Enzyme Interactions	189
		16.1.1 Electrophilic and Nucleophilic Groups	190
		16.1.2 Conjugation of Metabolite	190
	16.2	Interaction of Modulator with Metabolic Enzyme	191
		16.2.1 Enzyme–Substrate Interaction	192
		16.2.2 Enzyme Conformation	192

		16.3	Michael Acceptor Functionalities	192
		16.4	Enzyme Modulators with Michael Acceptor Characteristics	193
		16.5	Diversities of Enzyme Inducers	194
			16.5.1 Ortho-Hydroxyl Group on Aromatic Ring	194
			16.5.2 Chemical Structures of Enzyme Inducers	194
		Bibliography		201
17	**Metabolite-Associated Disease Conditions**			203
		17.1	Metabolite-Associated Hepatotoxicity	204
			17.1.1 Alcohol and Aldehyde	204
			17.1.2 Aflatoxin-Induced Hepatic Carcinogenesis	205
			17.1.3 Acetaminophen-Induced Hepatocyte Injury	205
			17.1.4 Other Factors	206
		17.2	Metabolic Intermediate–Associated Kidney Toxicities	206
		17.3	Metabolite–Associated Cancer and Other Toxicities	207
			17.3.1 Cancer	207
			17.3.2 Neurodegeneration	208
			17.3.3 Cataract	208
		17.4	Drug Efficacy and Adverse Responses	208
			17.4.1 Drug Efficacy	209
			17.4.2 Drug Adverse Responses	209
		17.5	Chemoprevention Against Toxicities	210
			17.5.1 Toxic Metabolites	210
			17.5.2 Chemoprevention	210
			17.5.3 Induction of Detoxification Enzymes	210
			17.5.4 Nrf2- Keap1 Pathway	211
		Bibliography		211
18	**Metabolic Enzyme Induction for Health Benefits**			215
		18.1	Metabolic Enzyme Modulation	216
			18.1.1 Activation Enzyme Modulation	216
			18.1.2 Detoxification Enzyme Modulation	218
			18.1.3 Balance Between Activation and Detoxification Inductions	219
		18.2	Varieties of Metabolic Enzyme Inducers	220
			18.2.1 Activation Enzyme Inducers	221
			18.2.2 Detoxification Enzyme Inducers	222
		18.3	Monofunctional and Bifunctional Inducers	223
			18.3.1 Monofunctional Inducers	223
			18.3.2 Bifunctional Inducers	224
		18.4	Inducer–Drug Interactions	224
		Bibliography		225
19	**Diet Effects on Metabolic Enzymes**			229
		19.1	Dietary Modulation of Metabolic Enzymes	230
			19.1.1 Vegetables	230
			19.1.2 Cruciferous Vegetables	230

		19.1.3	Allium Vegetables	233
		19.1.4	Root Vegetables	233
	19.2	Fruits		235
		19.2.1	Polyphenols	235
		19.2.2	Eriodyctiol and Quercetin	235
		19.2.3	Kaempferol and Pomegranate	236
		19.2.4	Anthocyanin and Procyanidin	236
		19.2.5	Triterpenes	236
	19.3	Herbs		237
		19.3.1	Ginseng	237
		19.3.2	Herb–Drug Interaction	237
	19.4	Beverage		237
		19.4.1	Epigallocatechin-3-Gallate	237
		19.4.2	Polyphenols	238
	19.5	Alcohol		238
		19.5.1	Acetaldehyde	238
		19.5.2	Curcumin Protective Effect	238
	19.6	Algae		239
		19.6.1	Chlorophyll	239
		19.6.2	Unsaturated Fatty Acid	239
		19.6.3	Polysaccharide	239
	19.7	Dietary Inducer–Drug Interactions		239
		19.7.1	Grapefruit	240
		19.7.2	Sulforaphane	240
		19.7.3	Curcumin	240
	Bibliography			240
Index				245

Overview

Foods are substances that contain essential nutrients such as proteins, carbohydrates, and minerals to provide energy and maintain life. Chemical compounds are a major part of dairy and pharmaceutical industries. Household products containing chemical compounds are used daily. Industrial chemicals and environmental pollutants are present in the air and rivers. Humans are constantly exposed to these foreign compounds (xenobiotics).

A large number of foreign compounds that find their way into the body are lipophilic (fat soluble) in nature. Unlike hydrophilic substances that are soluble in water, lipophilic compounds are nonpolar, insoluble, or have low solubility in water, which require conversion into water soluble compounds before they are eliminated and excreted from the body. The elimination of foreign compounds to which humans are exposed involves activation and detoxification mechanisms, which are important defenses that help humans.

During metabolism, some foreign compounds undergo metabolic conversion to metabolic reactive intermediates. A high expression of the activation enzyme may give rise to an overload of reactive chemical intermediates or metabolites. An extraordinary low efficiency of the detoxification enzyme may result in an abnormal detoxification metabolism. Maintaining metabolic reactive intermediates at a minimum level depends on the relative efficacies of activation enzymes and detoxification enzymes and a delicate balance between the reactions catalyzed by these enzymes.

The body's major defense mechanism against metabolic chemical intermediates is to minimize their exposure by speedily removing them from the body. To achieve this goal, the body develops metabolic enzyme systems. The expressions of activation and detoxification enzymes may vary among individuals. Variations in enzyme genetic polymorphisms as well as environmental and lifestyle factors can affect an individual's susceptibility to foreign compound-mediated toxic effects.

1.1 Foreign Compounds That Humans Expose

Foreign compounds that humans expose include foods, drugs, household products, environmental chemicals, and lifestyle substances. Foods include heterocyclic amines, nitrosamines, polycyclic aromatic hydrocarbons, azo dyes, unsaturated aldehydes, and mycotoxin; drugs include acetaminophen, xanthine, terfenadine, menadione, and diazepam; household products are such as benzene, phenol, and phthalate; environmental chemicals include diesel exhaust, arsenic, polychlorinated biphenyl, and dioxins; and lifestyle substances are such as cigarettes and alcohol.

1.2 Metabolic Reactions of Foreign Compounds

Phase I activation enzymes catalyze a variety of reactions to introduce functional groups to foreign compounds. Such varieties of reactions take place at specific atoms or groups. While phase II detoxification enzymes catalyze various conjugation reactions to form conjugated compounds that facilitate the excretion of foreign compounds, knowledge about these catalytic reactions and their functional characteristics is fundamental to understanding how these enzymes act on foreign compounds and what foreign compounds are metabolized by certain enzymes.

1.3 Activation Enzymes

Phase I activation enzymes catalyze the conversion reaction called functionalization for a lipophilic foreign compound, where a functional group is introduced to its chemical structure through oxidation, hydrolysis, or a reduction reaction. Functionalization increases the polarity of a lipophilic foreign compound, making it ready for the next detoxification metabolic step. Phase I enzymes are therefore referred to as activation enzymes. A hydrophilic foreign compound that has a functional polar group may bypass such a functionalization process.

1.4 Detoxification Enzymes

A functionalized lipophilic compound or a hydrophilic compound with a functional polar group is further metabolized by the phase II reactions. Phase II detoxification enzymes catalyze conjugation or non-conjugated reactions. In conjugation reaction, the functional group of a foreign compound is combined with a chemical group of a small molecule, leading to increases in the solubility of foreign compounds and thus facilitating their removal from the body.

Phase II enzymes are therefore referred to as detoxification enzymes. Detoxification of a foreign compound is an effective mechanism for the body's defense against foreign compound-mediated toxic effects. The major site of detoxification processes is the liver, but many metabolic reactions also occur in other organs such as the lung, kidney, and intestine.

1.5 Reactive Metabolic intermediates

Although xenobiotic metabolic pathways are beneficial to living organisms, however, in many cases, the generated metabolic intermediate or metabolite is acutely or potentially toxic. Furthermore, a significant number of reactive intermediates have the potential to react with oxygen to form reactive oxygen species and free radicals. Besides antioxidant enzymes, the body also relies on metabolic reactions catalyzed by detoxification enzymes to reduce and remove such reactive chemical intermediates and free radicals.

1.6 Oxidative Stress, Electrophiles, and Free Radicals

Oxidative stress occurs as a result of an imbalance between the production of reactive chemical species derived from foreign compound metabolism and the ability of organisms to neutralize such reactive chemical species. While inflammation is a biological response to oxidative stress, oxidative stress and chronic inflammation play a significant role in the initiation of pathological mechanisms underlying many disease conditions.

Electrophiles are electron-deficient species. They accept an electron pair from electron-rich species, such as carbocations and carbonyl compounds. Electrophiles are often positively charged molecules that do not have an octet of electrons. While free radicals formed in metabolic activation include superoxide anion radical and hydroxyl radical, free radical species can bind covalently to cellular macromolecules (proteins, nucleic acids, and lipids) and can also promote lipid peroxidation in cellular membranes.

1.7 Lifestyle Modifications

Lifestyle modifications include alcohol and cigarettes. Alcohol is hepatotoxic through metabolic disturbances associated with the oxidation of ethanol. Induction of microsomal enzymes results in increased acetaldehyde generation from ethanol. Acetaldehyde promotes glutamyl-cysteinyl-glycine depletion, free radical-mediated toxicity, and lipid peroxidation.

Cigarette smoke is a well-established risk factor because it contains toxic reactive molecules that are able to induce oxidative stress. Tobacco-specific nitrosamines are a group of carcinogens that are present in tobacco smoke. They contribute to smoking-related diseases, including cardiovascular and pulmonary diseases.

1.8　Metabolic Intermediate-Induced Cell Toxicities

Major targets for oxidative damage are cellular molecules, including proteins, nucleic acids, and lipids. Xenobiotic-induced oxidative protein damage can cause an impairment of enzyme catalytic function or other protein functions. Oxidative DNA damage affects DNA base pairing, causing a mismatch in the DNA base pair transformation and resulting in a mutation.

Oxidation of cell membranes leads to the peroxidation of a membrane lipid, which is a primary mechanism of tissue injury. Reactive intermediates produced during oxidative stress are believed to contribute to the development and progression of a variety of age - related diseases and other disease conditions.

1.9　Genetic Polymorphism of Metabolic Enzymes

Studies of individual responsiveness to foreign compounds have revealed considerable deviation, partly due to variations in their metabolisms. Among these variants is the difference in the level of expression of foreign compound-metabolizing enzymes, which may result in the observed variations in the potency of metabolic intermediates or metabolites.

The distinctive susceptibility of some individuals to their potentially toxic effects caused by metabolic enzyme genetic polymorphisms is an important factor in attributing to individual variations in the efficacy of activation and detoxification enzymes, especially cytochrome P450 and glutathione-S-transferase.

1.10　Defenses Against Metabolic Intermediates

To achieve speedy removal of xenobiotics, the body develops a number of enzyme systems that catalyze the conversion of lipophilic compounds to water-soluble hydrophilic metabolites. As discussed above, two distinctive steps in overall metabolic processes are phase I and phase II metabolisms, and their enzymes are referred to as phase I enzymes and phase II enzymes, respectively. These enzymes are produced from information stored in the genes.

1.10.1　Inducibility of Metabolic Enzymes

An important toxicologically relevant feature associated with activation and detoxification enzymes is the potential of these enzymes for induction or inhibition by some diaries or chemical compounds. The inducibility of these enzymes makes it possible to modify their expressions for health benefits.

1.10.2 Diversified Classes of Enzyme Modulators

Foreign compound metabolisms require substrate-enzyme interactions involving the reactive group of metabolites and the function group in enzyme amino side chains. The activities of foreign compound metabolic enzymes interfere with substrate-enzyme interactions through either covalently or non-covalently interactions.

Known compounds of importance in the modulation of foreign compound metabolic enzymes include (a) sulforaphane and other isothiocyanates, (b) 1,2-dithiole-3-thione and derivatives, (c) indole-3-carbinol, (d) flavonoids and isoflavones, (e) polyphenols, (f) organosulfur compounds, and (g) terpenes and terpenoids. These major metabolic enzyme inducers and their effects on metabolic enzyme modulation are also discussed in this book.

1.10.3 Induction and Inhibition Compound

As mentioned above, metabolic enzymes have the potential to be inducted or inhibited by some chemical compounds. Induction or inhibition of metabolic activation and detoxification enzymes has a significant impact on the extent of the toxicities of xenobiotics. The effects of metabolic enzymes on xenobiotic toxicity depend on not only the nature of the foreign compounds, but also the relative activities of these enzymes.

The expression of metabolic enzymes can also differ significantly as a result of exposure to environmental chemicals or dietary inducers. Strategies for protecting cells from initiation toxic events include a decrease in the expression of activation enzymes responsible for the generation of reactive species as well as an increase in the activities of detoxification enzymes that detoxify reactive intermediate species that intervene in cellular processes.

1.11 Defense Mechanisms: Nrf2-ARE Pathway

The activity of the antioxidant response pathway is largely regulated by nuclear factor erythroid-2-related factor 2 (Nrf2). The Nrf2 activation mechanism occurs as Nrf2 dissociates from its inactive complex with the repressor protein Keap1 and subsequently translocates into the nucleus. Once in the nucleus, the Nrf2 complex binds to antioxidant response elements (ARE), promoting transcription of cytoprotective genes.

1.12 Health Effects of Metabolic Intermediates

Chemical-induced liver injuries are caused by reactive intermediate species. Environmental and lifestyle factors also affect the susceptibility of individuals to such metabolite-induced injuries. The pathogenesis of most chemical-induced hepatotoxicity is initiated by electrophilic compound or free radical-induced oxidative stress. Selective toxicity to the kidney is due to its capability to accumulate intermediates and bioactivate these intermediates into toxic metabolites.

Cancer cells were also found to exhibit higher basal levels of reactive oxygen species in comparison with normal cells. Oxidative stress is linked to the development of neural dysfunction, which suggests a pathogenic role of oxidative stress in Alzheimer and Parkinson diseases. Moreover, xenobiotic - metabolic enzymes play a significant role in adverse drug response, because many drugs that are associated with adverse responses are subjected to metabolism by xenobiotic metabolic enzymes.

1.13 Induction of Metabolic Enzymes for Health Benefits

Modulation of metabolic enzymes affects the generation of reactive intermediates. Such modulation of metabolic enzymes to reduce foreign compound-mediated toxic effects has been a subject of intense interest. In general, the induction of activation enzymes can lead to a higher expression of enzyme activity and result in enhancements of the activation rate and the production of potentially toxic metabolic intermediates, while the induction of detoxification enzymes can lead to a higher expression of detoxification activity and an increase in the rate of detoxifying reactions.

However, the consequence of enzyme modulation is dependent on not only the specific effect on activation or detoxification enzymes but also the balance between the activities of activation and detoxification enzymes. These subjects are also discussed in detail in the book.

1.14 Dietary Effects on Metabolic Enzymes

Dietary effects on the human health have been a subject of intensive studies. Advances in the understanding of mechanisms that govern the detoxification of foreign compounds have revealed that diets can have important impacts on the efficacy of activation and detoxification enzymes. Many chemical compounds that are capable of acting as enzyme modulators are present in the daily human diet. Extensive research has been carried out to explore how these enzymes can be modulated in human diets for health benefits.

An increase in the intake of diets rich in inducers of detoxification enzymes is considered a promising proposal to minimize foreign compound-mediated toxic effects. Diets rich in vegetables and fruits that contain enzyme modulators have received much attention, in particular those rich in inhibitors of activation enzymes, such as cytochrome P450, and those rich in inducers of detoxification enzymes, such as uridine-diphosphate-glucuronosyl-transferase and glutathione S-transferase.

Bibliography

Boelsterli UA (2007) Mechanistic toxicology. CRC Press, Boca Raton
Buxton ILO, Benet LZ (2011) Pharmacokinetics: the dynamics of drug absorption, distribution, metabolism and elimination. In: Brunton LL et al (eds) Goodman & Gilman's the pharmacological basis of therapeutics. McGraw-Hill, New York
Chen C-H (2012) Activation and detoxification enzymes: functions and implications. Springer Sciences, New York
Chen C-H (2020) Xenobiotic metabolic enzymes: bioactivation and antioxidant defense. Springer Nature, Switzerland
Conney AH (2003) Enzyme induction and dietary chemicals as approaches to cancer chemoprevention. Cancer Res 63:7005–7031
Finley JW, Schwass DE (1985) Xenobiotic metabolism: nutritional effects. American Chemical Society, Washington, DC
Giacomini KM, Sugiyama Y (2011) Membrane transports and drug response. In: Brunton LL et al (eds) Goodman & Gilman's the pharmacological basis of therapeutics. McGraw-Hill, New York
Gonzalez FJ, Coughtrie M (2011) Drug metabolism. In: Brunton LL et al (eds) Goodman & Gilman's the pharmacological basis of therapeutics. McGraw-Hill, New York
Hodgson E, Das PC, Cho TM, Rose RL (2008) Phase I metabolism of toxicants and metabolic interactions. In: Smart RC, Hodgson E (eds) Molecular and biochemical toxicology. Wiley, New York
Ioannides C (2002) Xenobiotic metabolism: an overview. In: Ioannides C (ed) Enzyme systems that metabolise drugs and other xenobiotics. Wiley, New York
Jakoby WB (1980) Enzymatic basis of detoxication. Academic, New York, pp v1–v2
Jakoby WB, Bend JR, Caldwell J (1982) Metabolic basis of detoxication. Academic, New York
Josephy PD, Mannervik B, de Montellano PO (1997) Molecular toxicology. Oxford University, New York
LeBlanc GA (2008) Phase II-conjugation of toxicants. In: Smart RC, Hodgson E (eds) Molecular and biochemical toxicology. Wiley, New York
Lee JS, Obach RS (2003) Drug metabolizing enzymes. Marcel Dekker, New York
Mulder GJ (1990) Conjugation reactions in drug metabolism: an integrated approach. Taylor and Francis, London
Parkinson A, Ogilvie BW (2008) Biotransformation of xenobiotics. In: Klaassen CD (ed) Casarett & Doull's toxicology: the basic science of poisons. McGraw-Hill, New York
Sardesai VM (2003) Introduction to clinical nutrition. Marcel Dekker, New York

Foreign Compounds: Foods, Drugs, Chemicals, and Lifestyles

Foods are substances that contain essential nutrients such as proteins, carbohydrates, and minerals to provide energy and maintain life. Drugs are chemical substances that produce biological effects that affect how the body works. A medicine is used in the treatment, diagnosis, or prevention of a disease. Chemical compounds are a major part of the dairy and pharmaceutical industries. They are also used for preservation in the food industry. Fertilizer chemicals are applied to accelerate the growth of crops, while alcohol and cigarettes are lifestyle substances.

Typical foreign compounds, including foods, drugs, household products, environmental chemicals, and lifestyle substances, are presented in Fig. 2.1.

2.1 Foods

Vegetables and fruits that humans ingest are largely natural plants. A diverse group of chemicals are present in grains and fruits. Natural plants generate a variety of biologically active chemicals to protect themselves. After their indigestion and metabolism, metabolites are generated via complex biochemical reactions that occur within the cells.

Contaminants in food can occur, including pesticides from crop sprays, fungi from storage, phthalate ester from packaging, and styrene from containers. In addition, residues of antibiotics and hormones used to raise chickens, cattle, pigs, and sheep may remain as contaminants in meat. Fish may be contaminated with industrial wastes such as mercury, PCBs, and dioxins. Chemical derivatives are produced when meat or fish is cooked at high temperatures.

Besides, molds may produce metabolites with the potential to produce adverse health effects.

Known potentially harmful chemicals in foods include heterocyclic amines, nitrosamines, polycyclic aromatic hydrocarbons, azo dyes, unsaturated aldehydes and mycotoxins, which are discussed below:

Foods	Drugs	Household Products	Environmental Chemicals	Lifestyles Substances
Heterocyclic amines	Acetaminophen	Benzene	Diesel exhaust	Cigarette
Nitrosamines	Xanthine	Phenol	Arsenic	Alcohol
Polycyclic aromatic hydrocarbons	Terfenadine	Phthalate	Polychlorinated biphenyl	
Azo dyes	Menadione		Dioxins	
α,β-unsaturated aldehydes	Diazepam			
Mycotoxin				

Fig. 2.1 Typical foreign compounds that humans are exposed

2.1.1 Heterocyclic Amines

Heterocyclic amines are organic compounds that contain at least one atom of carbon and at least one atom of nitrogen within aromatic or non-aromatic rings, such as pyridine or pyrimidine. More than a dozen heterocyclic amines have been identified. 2-amino-1-methyl-6-phenylimidazo [4,5-beta] pyridine is the most abundant heterocyclic amine found in human diets.

As meats or fish are fried or barbecued, heterocyclic amines are formed when amino acids react with creatine in muscle. Many heterocyclic amines have been reported to be carcinogenic. They are metabolized by activation enzymes such as cytochrome P450. Chronic administration of these chemicals was found to induce tumors in rats.

2.1.2 Nitrosamines

Nitrosamines containing an organic functional group N-N=O are formed by the addition of N=O group to secondary or tertiary amines. N-nitroso derivatives of amines are formed by reaction between nitrite and amines. Nitrosamines are found in foods such as mushrooms, fermented and smoked fish, and pickled foods. Cured meats, such as bacon, can also contain nitrosamines because sodium nitrite is added as a preservative.

High cooking temperatures, such as fry bacon, also contribute to nitrosamine formation. Nitrosamines may also be found in gastric juice, possibly formed by reaction between amines and nitrites from the diet. In addition, tobacco-specific nitrosamines are generated during the fermentation and burning of the tobacco leaf.

Chemical carcinogens generally require metabolic activation in order to bind to DNA, leading to causing mutation. Many nitrosamines have been reported to be carcinogenic in experimental animals. For example, N-nitrosodibutylamine and its hydroxylated metabolite are urinary bladder-specific carcinogens. Evidences also support the role of tobacco-specific nitrosamines as an important contributing factor for cancers in humans.

Human susceptibility to nitrosamine toxicity varies from individual to individual, depending on the expression of metabolic enzymes. Nitrosonornicotine was found to significantly increase activation enzyme expression in the liver and the lung. It also noticeably decreases the expression of glutathione levels and glutathione S-transferase.

2.1.3 Polycyclic Aromatic Hydrocarbons

Polycyclic aromatic hydrocarbons are products of incomplete combustion at high temperatures. Major polycyclic aromatic hydrocarbons detected in charcoal meats include benzopyrene and dibenzanthracene. Hence, humans could be exposed to polycyclic aromatic hydrocarbons by consuming grilled or charred meats.

Polycyclic aromatic hydrocarbons are metabolically activated by activation enzymes such as cytochrome P450 (CYP450). Aromatic hydrocarbons exhibit their toxic properties after metabolic conversion to reactive chemical intermediates. For example, following absorption, dibenzanthracene is distributed to various tissues, with the highest accumulation in the liver and kidneys.

Benzopyrene is also a major lung carcinogen, which toxicity is produced by the bioactivation of a toxic intermediate by CYP450. Benzopyrene has the capacity to interact with DNA. A considerable number of studies have documented the link between benzopyrene and cancer. Dibenzanthracene was also shown to induce hepatic aryl hydrocarbon hydroxylase activity in mice. Regular consumption of overcooked charcoal barbecued beef has also been reported to be associated with increased levels of colon cancer.

2.1.4 Azo Dyes

Azo dyes are stable in the pH ranges of food products. Their colors do not fade when exposed to light or oxygen. Such properties make azo dyes applicable for use in the food industries. Many azo pigments are non-toxic. Hence, the acute toxicity is low in consuming azo dye-colored foods. However, some azo dyes have been banned for food use because of their side effects due to their degradation products.

Azo dyes undergo enzymatic breakdown catalyzed by azo-reductase present in various microorganisms. Azo-reductase activity is high in the liver and kidney. After cleavage of the azo-linkage, amine metabolites are absorbed in the intestine and excreted in the urine. Azo degradation products have been reported to be mutagenic or carcinogenic. Enzymatic sulfonation appears to decrease the toxicity of azo degradation products by facilitating urinary excretion of azo dyes and their metabolites.

2.1.5 α,β-Unsaturated Aldehydes

Cooking oils that are used in caterings and households can generate significant amounts of oxygenated α,β-unsaturated aldehydes. For example, 4-hydroxynonenal is an α,β-unsaturated hydroxyalkenal generated in the oxidation of lipids. It is present in higher quantities as the lipid peroxidation chain reaction increases. Through diets, the body absorbs 4-hydroxynonenal and oxygenated α,β-unsaturated aldehydes. These compounds are considered potential contributing agents to a number of disease conditions, such as inflammation, respiratory distress syndrome, and diabetes.

2.1.6 Mycotoxin

Mycotoxin is a toxin produced by the fungi of mushrooms, molds, or yeasts. When fungi propagate into colonies, mycotoxin levels become high. Wild mushrooms contain an assortment of mycotoxins that can cause health problems. Mycotoxin appears in food chains as a result of fungal infection of crops, which can also remain in the food chains of meat and dairy products.

Patulin, a mycotoxin produced by a variety of molds, is commonly found in rotten apples. Studies of rat liver tissues reported that patulin decreases glutathione S-transferase activity and markedly increases lipid peroxidation. Such effects may be a result of a patulin-mediated reduction in the level of glutathione. Decreases in glutathione levels and glutathione S-transferase activity may be related to the presumed mutagenic or carcinogenic potential of patulin.

Besides, poultry is susceptible to the toxic effects of aflatoxin B_1, another mycotoxin, due to the results of effective activation by CYP450 and deficient detoxification by glutathione S-transferases. Aflatoxin B_1 can cause health problems by acting as an allergen or irritant or by weakening the immune system.

2.2 Household Products

Household products, including lubricants, detergents, paints, and pesticides, contain a variety of organic solvents. Humans are exposed to many organic solvents through inhalation, including gasoline, paint removers, varnishes and wood sealants. In industries, organic solvents are used to make resins, nylon, synthetic fibers, and plastics. Volatile organic solvents are also present in dry cleaning shops, electronics industries, and scientific laboratories. Benzene and phenol are typical examples of organic solvents present in household products.

2.2.1 Benzene

Benzene is present in gasoline and cigarette smoke. Exposure to benzene is known to cause human health problems. The toxic effects of benzene in humans are attributed to hydroxylated metabolites, such as hydroquinone and phenol. Long-term exposure of benzene can have harmful effects on the bone marrow and can cause a decrease in red blood cells, leading to anemia. Benzene was also found to affect the immune system, causing increases in the chance of infection, depression of the immune system, and cancer in humans.

2.2.2 Phenol

Phenol is the primary metabolite of benzene. Detoxification enzyme-catalyzed conjugation reactions are involved in detoxifying phenol. At a low exposure concentration of benzene, phenylsulfonate is the major conjugate of phenol in the blood. However, at a high exposure concentration, phenylglucuronide is the predominant conjugate. Reductions in spleen weight and white blood cell numbers were reported to correlate with the concentration of phenylsulfonate in the blood.

2.2.3 Phthalate

Phthalates are one of the most widely used plasticizers in polymer products. They are also used in capsules and enteric coatings. Their exposure may generate health concerns. Phthalates are among environmental contaminants and endocrine-disrupting chemicals. An association between phthalate exposure and allergic diseases has been suggested.

2.3 Pharmaceuticals

Active ingredients in drugs perform the needed actions for therapeutic responses. The time course and duration of action are affected by the enzymatic activities associated with the metabolisms of these drugs. When the active constituent of a drug is metabolized into a reactive chemical intermediate or metabolite, drug-mediated toxic effects occur if detoxification enzymes are inefficient in eliminating it from the body.

The majority of medications that humans take are synthetic chemical compounds. On taking specific drugs, individuals may experience some side effects derived from drug metabolism. This can occur with drugs such as painkillers and diuretic. For instances, acetaminophen is a major ingredient in painkillers. An overdose of acetaminophen can cause damage to the liver. Furosemide, a diuretic, if given in excessive amounts, can lead to profound diuresis with a depletion of water and electrolytes.

Overdoses of drugs are a health problem. Harmful effects can be attributed to the accumulation of drug ingredients in the body. A low expression of detoxification enzymes to metabolize these drugs is also a major concern. Examples of potential overdose of drugs include acetaminophen, xanthine, terfenadine, menadione, and diazepam.

2.3.1 Acetaminophen

Acetaminophen is one of the most common pharmaceutical agents that are involved in overdose and toxicity. Its misuse or overdose is associated with hepatotoxicity. This drug is metabolized by the activation enzyme CYP450, while glucuronide and sulfonate conjugates are involved in the metabolism of acetaminophen detoxification.

In the liver, acetaminophen is metabolized by a conjugation reaction to form water-soluble conjugate for elimination in the urine. However, a deficiency in a conjugation enzyme, such as UDP-glucuronosyl transferase, can cause the accumulation of reactive chemical intermediate or metabolite that is capable of interacting with cellular proteins and membranes, resulting in hepatocellular damages.

2.3.2 Xanthine

Bronchodilators xanthine and its derivatives are a group of alkaloids that are commonly used as mild stimulants and in the treatment of the symptoms of asthma. Xanthine is a product of the pathway of purine degradation. Metabolic conversion of xanthine to uric acid, carried out by an enzymatic reaction catalyzed by xanthine oxidase, leads to subsequent excretion of xanthine from the body. In the absence of sufficient xanthine oxidase, xanthine cannot be readily converted to uric acid. An accumulation of xanthine in the body could result in xanthine-mediated oxidative stress.

2.3.3 Terfenadine

Terfenadine, an antihistamine, is metabolized by CYP450 to its metabolite fexofenadine. This metabolism is blocked by drugs such as erythromycin or foods such as grapefruit, making it difficult to metabolize and remove from the body. An elevated level of terfenadine can lead to an adverse cardiac effect on the heart rhythm. Potential toxicity can also occur as a result of its interaction with other medications.

2.3.4 Menadione

Menadione is one of the man-made versions of vitamin K. Menadione is a lipophilic vitamin precursor that is converted into menaquinone in the liver. A deficiency in vitamin K, especially in infants, can be a severe health issue because they easily suffer from extensive hemorrhaging. While an overdose of menadione can be equally detrimental. Overdoses of menadione have been reported to cause adverse effects, including hemolytic anemia and neonatal brain or liver damage. Menadione supplements were banned because of their potential toxicity in human use.

2.3.5 Diazepam

Diazepam belongs to a class of drugs known as benzodiazepines. It is used to treat anxiety, alcohol withdrawal, and seizures. Studies were conducted to ascertain any involvement of free radical-mediated prooxidative processes in brain regions following diazepam administration. A significant decrease in glutathione reductase activity was also observed.

2.4 Environmental Chemicals

Industrial combustions in refineries, incineration, and coal plants produce industrial pollutions, including polycyclic aromatic hydrocarbons, dioxins, and polychlorinated biphenyls. Transportation vehicles generate exhaust gases. Known environmental toxicants that humans expose also include arsenic and polychlorinated biphenyls present in contaminated drinking water.

Such environmental toxicants to which humans are exposed are of major concern. Their potentially harmful compounds are either intrinsically toxic or become toxic after metabolic conversion. They have been reported to play an important role in the pathogenesis of lung disease and other disease conditions, such as cancer.

2.4.1 Diesel Exhausts

Diesel exhausts produced in the combustion of diesel fuel are a mixture of gases and fine particles that contain harmful air contaminants, such as polycyclic aromatic hydrocarbons. Diesel exhausts also include many potential cancer-causing substances such as benzene, arsenic, and formaldehyde, as well as other toxic pollutants such as nitrogen oxides. Diesel exhaust particles can also initiate allergic responses.

Exposure to diesel exhausts can cause inflammation in the lungs, aggravating chronic respiratory symptoms and increasing the frequency or intensity of asthma attacks. Such harmful effects are especially pronounced in individuals whose detoxification enzyme expression is impaired. Chemically reactive intermediates are believed to play a key role in cellular damage after exposure to diesel exhaust particles. Glutathione S-transferase was reported to be involved in the detoxification of diesel exhaust particle-mediated allergic inflammation.

2.4.2 Arsenic

Arsenic is present in polluted drinking water. Populations in many regions of the world are chronically exposed to arsenic-contaminated drinking water. When rice is grown in polluted water, paddy rice takes up arsenite from the soaking soil. Groundwater pollution by arsenic is a serious worldwide problem. Drinking water from arsenic-tainted wells causes ailments marked by rough skin and often leads to serious diseases such as skin or bladder cancer.

Arsenic is metabolized to monomethylarsonic acid, which is then converted to dimethyarsinic acid by methyltransferase enzymes. The arsenic metabolite requires S-adenosyl-methionine as the methyl donating cofactor, before its excretion through urine. Arsenic was also reported to interfere with methyltransferases and inactivate tumor suppressor genes. Other studies reported that arsenic-induced malignant transformations are linked to DNA hypomethylation.

2.4.3 Polychlorinated Biphenyls

Polychlorinated biphenyls (PCBs), a major class of persistent organic pollutants, are a group of man-made organic chemicals that consist of carbon, hydrogen, and chlorine atoms. Due to their non-flammability and chemical stability, PCBs are used in a variety of industrial and commercial applications. However, PCBs do not readily break down and can remain for long periods in water and soil.

PCBs are metabolized to hydroxylated compounds. Many of these metabolites are further converted to either sulfonate or glucuronic acid conjugates catalyzed by detoxification enzymes such as sulfotransferase or uridine diphosphate glucuronosyl transferase. PCBs were also reported to induce the activity of monooxygenases that catalyze the metabolism of PCBs, leading to the formation of metabolites and potential adverse health effects.

Studies in humans support evidence for the potential carcinogenic and non-carcinogenic effects of PCBs. They have been shown to cause cancer in animals as well as a number of serious non-cancer health effects in animals, including the effects on the immune, reproductive, nervous, and endocrine systems.

2.4.4 Dioxins

Dioxins are mainly by-products of industrial processes, including smelting, chlorine bleaching of paper pulp, and manufacturing herbicides and pesticides. Dioxins can also result from natural processes such as volcanic eruptions and forest fires. They are environmental pollutants. Dioxins exert their effects through interaction with an intracellular protein, the Ah receptor. The Ah receptor functions as a transcriptional enhancer that interacts with a number of other regulatory proteins.

Dioxins were also reported to modulate interactions with specific base sequences in the DNA. Short-term exposure to high levels of dioxins may result in skin lesions and liver function alteration. Long-term exposure is linked to impairments of the immune, nervous, and endocrine systems, as well as reproductive functions. Immunotoxicity and effects on the reproductive system appear to be among the most sensitive responses to dioxins.

2.5 Lifestyles

Common foreign lifestyle compounds are alcohol and cigarettes. Ethanol is absorbed by the digestive tract, mainly in the small intestine. Alcohol is then metabolized mainly in the liver, where it is converted into acetaldehyde. Acetaldehyde is then metabolized to acetate and converted into acetyl-CoA in other tissues.

Cigarette smoke contains a great variety of chemicals, many of which are toxic. Among them are nitrosamines and polycyclic aromatic hydrocarbons. The association of cigarette smoke with a higher level of chronic inflammation and other disease conditions is well documented.

2.5.1 Alcohol

Extensive research has been carried out to examine the toxic and beneficial effects of ethanol and its metabolites (acetaldehyde). The toxic effects are based on high alcohol consumption, while the beneficial effects are focused on low to moderate consumption. Oxidative metabolisms by alcohol dehydrogenase, cytochrome P450 2E1, and aldehyde dehydrogenase were reported to play important roles in the effects of alcohol.

Alcohol undergoes diffusion throughout the body. The heart and vascular system are susceptible to the effects of alcohol. Heavy alcohol intake increases the risk of hypertension. However, the relationship between light or moderate alcohol consumption and health effects remains controversial.

2.5.2 Cigarette

Humans are exposed to cigarette smoke and auto exhausts more commonly than other environmental sources. Typical toxic chemicals produced in cigarette smoke include nitrosamines, nicotine, and benzo(a)pyrene. Cigarette-generated nitrosamines are composed of various amines, such as nicotine. Nicotine is rapidly absorbed into the bloodstream and reaches the brain within 10 seconds. It affects the brain chemistry, resulting in a number of chemical reactions that involve hormones and neurotransmitters.

Two major carcinogens in cigarette smoke are benzo[a]pyrene and dibenz[a,h] anthracene. Benzo(a)pyrene in cigarette smoke resulting from the combustion of organic materials was found to have the capacity of binding to cell components in major organs. Polycyclic aromatic hydrocarbons present in cigarette smoke are due to the combustion of organic materials.

Bibliography

Balliet RM, Chen G, Dellinger RW et al (2010) UDP-glucuronosyltransferase 1A10: activity against the tobacco-specific nitrosamine, 4-(methylnitrosamino)-1-(3-pyridyl)-1-butanol, and a potential role for a novel UGT1A10 promoter deletion polymorphism in cancer susceptibility. Drug Metab Dispos 38:484–490

Bessems JG, Vermeulen NP (2001) Paracetamol (acetaminophen)-induced toxicity: molecular and biochemical mechanisms, analogues and protective approaches. Crit Rev Toxicol 31:55–138

Birnbaum LS (1994) The mechanism of dioxin toxicity: relationship to risk assessment. Environ Health Perspect 102(Suppl 9):157–167

Bolling AK, Sripada K, Becher R et al (2020) Phthalate exposure and allergic diseases: Review of epidemiological and experimental evidence. Environ Int 139:105706

Bu-Abbas A, Clifford MN, Ioannides C et al (1995) Stimulation of rat hepatic UDP- glucuronosyl transferase activity following treatment with green tea. Food Chem Toxicol 33:27–30

Cao Z, Li Y (2004) Potent induction of cellular antioxidants and phase 2 enzymes by resveratrol in cardiomyocytes: protection against oxidative and electrophilic injury. Eur J Pharmacol 489:39–48

Chen C-H (2012) Activation and detoxification enzymes: functions and implications. Springer Sciences, New York

Chen C-H (2020) Xenobiotic metabolic enzymes: bioactivation and antioxidant defense. Springer Nature, Cham

Conaway CC, Wang CX, Pittman B et al (2005) Phenethyl isothiocyanate and sulforaphane and their N-acetylcysteine conjugates inhibit malignant progression of lung adenomas induced by tobacco carcinogens in A/J mice. Cancer Res 65:8548–8557

Dare BL, Lagente V, Gicquel T (2019) Ethanol and its metabolites: update on toxicity, benefits, and focus on immunomodulatory effects. Drug Metab Rev 51(4):545–561

Dashwood RH, Xu M, Hernaez JF et al (1999) Cancer chemopreventive mechanisms of tea against heterocyclic amine mutagens from cooked meat. Proc Soc Exp Biol Med 220:239–243

Elbekai RH, El-Kadi AO (2004) Modulation of aryl hydrocarbon receptor-regulated gene expression by arsenite, cadmium, and chromium. Toxicology 202:249–269

Furst A (2002) Can nutrition affect chemical toxicity? Int J Toxicol 21:419–424

Gooderham NJ, Murray S, Lynch AM et al (2001) Food-derived heterocyclic amine mutagens: variable metabolism and significance to humans. Drug Metab Dispos 29:529 534

Kensler TW, Groopman JD, Eaton DL et al (1992) Potent inhibition of aflatoxin-induced hepatic tumorigenesis by the monofunctional enzyme inducer 1,2-dithiole-3-thione. Carcinogenesis 13: 95–100

Li X, Parkin S, Duffel MW et al (2010) An efficient approach to sulfate metabolites of polychlorinated biphenyls. Environ Int 36:843–848

Macé K, Aguilar F, Wang JS et al (1997) Aflatoxin B1-induced DNA adduct formation and p53 mutations in CYP450-expressing human liver cell lines. Carcinogenesis 18:1291–1297

Maliakal PP, Coville PF, Wanwimolruk S (2002) Decreased hepatic drug metabolising enzyme activity in rats with nitrosamine-induced tumours. Drug Metabol Drug Interact 19:13–27

Manson MM, Ball HW, Barrett MC et al (1997) Mechanism of action of dietary chemoprotective agents in rat liver: induction of phase I and II drug metabolizing enzymes and aflatoxin B1 metabolism. Carcinogenesis 18:1729–1738

Medinsky MA, Kenyon EM, Schlosser PM (1995) Benzene: a case study in parent chemical and metabolite interactions. Toxicology 105:225–233

Musavi S, Kakkar P (1998) Diazepam induced early oxidative changes at the subcellular level in rat brain. Mol Cell Biochem 178(1–2):41–46

Paquot N (2019) The metabolism of alcohol. Rev Med Liege 74(5–6):265–267

Pegram RA, Chou MW (1989) Effect of nitro-substitution of environmental polycyclic aromatic hydrocarbons on activities of hepatic phase II enzymes in rats. Drug Chem Toxicol 12:313–326

Pfeiffer E, Diwald TT, Metzler M (2005) Patulin reduces glutathione level and enzyme activities in rat liver slices. Mol Nutr Food Res 49:329–336

Pool-Zobel B, Veeriah S, Böhmer FD (2005) Modulation of xenobiotic metabolising enzymes by anticarcinogens – focus on glutathione S-transferases and their role as targets of dietary chemoprevention in colorectal carcinogenesis. Mutat Res. 591:74–92

Shimada T (2006) Xenobiotic-metabolizing enzymes involved in activation and detoxification of carcinogenic polycyclic aromatic hydrocarbons. Drug Metab Pharmacokinet 21:257–276

Sree CG, Buddolla V, Lakshmi BA et al (2023) Phthalate toxicity mechanisms: an update. Comp Biochem Physiol C Toxicol Pharmacol 263:109498

Talalay P (1989) Mechanisms of induction of enzymes that protect against chemical carcinogenesis. Adv Enzyme Regul 28:237–250

Tampal N, Lehmler HJ, Espandiari P et al (2002) Glucuronidation of hydroxylated polychlorinated biphenyls (PCBs). Chem Res Toxicol 15:1259–1266

Thompson D, Oster G (1996) Terfenadine is indicated for the relief of symptoms associated with seasonal allergic rhinitis such as sneezing, rhinorrhea, pruritus, and lacrimation. J Am Med Assn 275:1339–1341

Vernhet L, Séité MP, Allain N et al (2001) Arsenic induces expression of the multidrug resistance-associated protein 2 (MRP2) gene in primary rat and human hepatocytes. J Pharmacol Exp Ther 298:234–239

Wang LQ, James MO (2006) Inhibition of sulfotransferases by xenobiotics. Curr Drug Metab 7:83–104

Wells MS (1991) Nerland DE. Hematotoxicity and concentration-dependent conjugation of phenol in mice following inhalation exposure to benzene. Toxicol Lett 56:159–166

Zhang H, Forman HJ (2009) Signaling pathways involved in phase II gene induction by alpha, beta-unsaturated aldehydes. Unsaturated aldehydes. Toxicol Ind Health. 25(4–5):269–278

Transport and Excretion of Foreign Compounds

Based on the solubilities in water, foreign compounds that humans ingest or inhale can be classified into two categories: hydrophiles (soluble in water) and lipophiles (soluble in lipid medium). Membrane lipid bilayers serve as physical barriers for transporting foreign compounds across cell membranes. Transport mechanisms for lipophilic and hydrophilic compounds are different. Being lipid-soluble, lipophilic molecules can generally diffuse across cell membranes. While hydrophilic substances are usually unable to penetrate lipid membranes, they require carriers or channels for crossing cell membranes due to their low lipid solubility.

The process by which foreign compounds cross cell membranes and enter the blood stream is referred to as absorption. Before they reach the blood stream, xenobiotics must cross the intestinal epithelium, basement membrane, and capillary endothelium. The gastrointestinal tract is one of the most important sites where foreign compounds are absorbed. When a foreign compound is absorbed, it undergoes metabolism in the liver before reaching the systemic circulation.

The urinary and biliary systems are two primary routes for excretion. Before being excreted from the body, lipophilic foreign compounds require metabolic conversion into hydrophilic metabolites. Various processes that foreign compounds go through before their excretion from the body are described in Fig. 3.1, which includes the entry, the absorption and the metabolic processes.

Due to lipid bilayers serving as physical barriers for biological membranes, hydrophilic foreign compounds cannot spontaneously enter the inner cell compartment. They are transported across cell membranes through transport mechanisms, such as facilitated diffusion and active transport. Passive diffusion, a major process for absorption, is dependent on the concentration gradient and the lipid-water partition coefficient of foreign compounds, while active transport requires acquiring energy to transfer foreign compounds against a concentration gradient across cell membranes.

Fig. 3.1 Absorption, metabolism, and excretion of foreign compounds

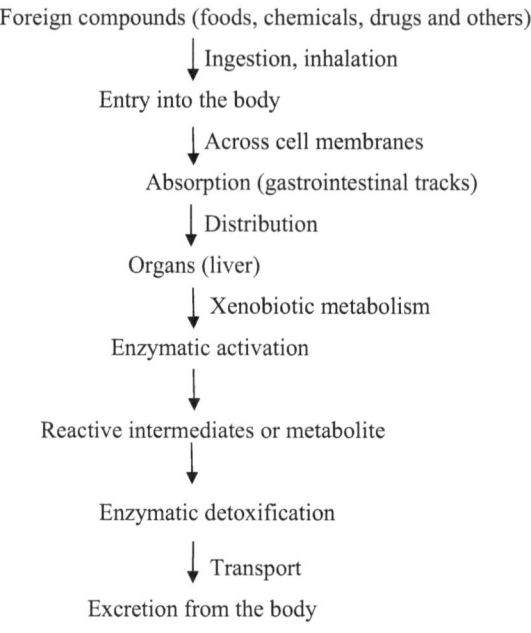

3.1 Lipophiles Versus Hydrophiles

Serving as the solvent for intracellular and extracellular media, water is a great solvent for ionic substances, partly due to its polarity. It is composed of a positive hydrogen end and a negative oxygen end, which gives a dielectric constant of 78. In comparison, the interior of biological membranes is a nonpolar medium with a low dielectric constant of around 4.

Table 3.1 presents some typical lipophilic foreign compounds that human ingest, including chemicals, drugs, and foods. Their relative solubility in water is also included in the table.

The polarity of water permits to solvate ionic compounds. Hydration occurs when solute compounds dissolve in water. Besides polarity, water also has the capability of forming intramolecular and intermolecular hydrogen bonding. The solubility of a foreign compound in water increases when it forms intermolecular hydrogen bonding with water.

The general rule of solute solubility in a solvent is "like dissolves like." That is, nonpolar or weakly polar compounds dissolve in nonpolar or weakly polar solvents, while polar compounds dissolve in polar solvents. Accordingly, lipophilic foreign compounds with low dielectric constants prefer to dissolve in low dielectric constant lipid mediums but not in high dielectric constant water solvent.

Table 3.1 Typical lipophilic foreign compounds

Foreign compounds	Solubility in water
Chemicals	
Benzo[a]pyrene	Insoluble
Quinone	Slightly
Polychlorinated biphenyls	Insoluble
Aniline	Insoluble
Azo	Insoluble
Dimethyl phthalate	Slightly
Drugs	
Menadione	Insoluble
Diazepam	Very slightly
Acetaminophen	Very slightly
Terfenadine	Slightly
Foods	
Turmeric	Low
Caffeine	Moderate soluble
Lycopene	Insoluble

3.2 Hydrogen Bonding

Water is a polar solvent. Based on the convention, like dissolves like, polar foreign compounds are soluble in water. The side chains of foreign compounds that form hydrogen bonding with water increase their solubility in water. Intermolecular hydrogen bonding with water enhances the solubility of a foreign compound.

3.3 Sites of Action

The accumulation of foreign compounds at the site of action is facilitated by absorption and distribution. Absorption is the transfer of a foreign compound from the site of exposure into the general circulation. Lipid solubility of foreign compounds is usually the most important character that influences their transport and absorption. While transporters may contribute to the gastrointestinal absorption of some chemicals, a large majority of foreign compounds traverse epithelial barriers and reach blood capillaries by diffusion through the cells.

While the metabolic site for drugs depends on the presence of metabolic enzymes, foreign compound metabolic enzymes are largely present in the liver. The liver encounters foreign compounds, such as chemicals, foods, and drugs, after they are absorbed in the intestinal tract. Thus, the liver is the major organ, where xenobiotics are metabolized, reactive intermediates are produced, and toxicity and carcinogenesis are manifested.

Hepatocytes contain phase I activation enzymes and phase II detoxification enzymes. Activation enzymes catalyze functionalization reactions to enhance the solubility of foreign compounds in water. Then, detoxification enzymes catalyze conjugation or non-conjugation reactions to detoxify foreign compounds and to facilitate their excretion from the body. Metabolic intermediates or metabolites may react at the site where they are generated.

The liver is not necessarily the target organ of toxicity. Before being transported to other organs, metabolic reactive intermediates or metabolites may potentially cause toxicity in the liver. The metabolites may also diffuse away and react with other targets. Metabolic intermediates may be transported to other organs, where they exert toxic effects.

Besides the liver, the kidney is a frequent target organ of toxicity and is also a main site for drug metabolism. The kidney, which receives a large amount of blood, also contains a variety of foreign compound-metabolizing enzymes. In addition, the breast, lung, and colon are frequent sites of tumorigenesis in spite of their limited metabolic ability. Moreover, metabolites of xenobiotics, such as aromatic amines, can also be transported to the bladder, where they are released and converted to carcinogenic species.

3.4 Cell Membranes

Living organisms use cell membranes as hydrophobic permeability barriers to control access to the internal cell compartment. Lipid bilayers serve as physical barriers that do not favor a spontaneous exchange of foreign compounds between the internal and external cell compartments. Accordingly, the physical barriers of cell membranes are attributed to the role of membrane permeability and hydrophobicity in the metabolism of foreign compounds.

Depending on their solubility characteristics, the movement of foreign compounds into or out of cell membranes is carried out by various transport mechanisms. Lipophilic, nonpolar compounds, are able to move across cell membranes. While hydrophilic, polar compounds are largely restricted to the extra-cellular compartment and cannot enter the cells simply by free diffusion. Consequently, the uptake of hydrophilic compounds across cell membranes is mediated by channels or transport proteins, which specifically select solutes from the extracellular medium.

3.5 Transport Mechanisms

The cell membrane consists of a lipid bilayer that regulates the entry and exit of foreign compounds. Major mechanisms for the transport of foreign compounds across biological membranes include passive diffusion, active transport, and facilitated diffusion. These transport mechanisms are discussed below.

3.5.1 Passive Diffusion

Passive diffusion of solutes across plasma membranes is composed of three steps: partition from the external aqueous medium to the membrane lipid phase, diffusion across membrane lipid bilayers, and partition into the internal cellular aqueous medium. In passive diffusion, the driving force for solute to move across membrane lipids into the cells is the concentration gradient, in which the concentration of solute in the external cell medium is higher than that in the internal cell medium.

3.5.2 Facilitated Diffusion

A large number of foreign compounds are also transported across cell membranes through facilitated diffusion. In facilitated diffusion, the transport of a foreign compound across membranes into the cells is facilitated by a transport protein (carrier). Facilitated diffusion does not require energy input. It occurs downhill in accordance with the solute concentration gradient.

Mechanisms of facilitated diffusion include channel, transporter, and active transport. They are briefly described below:

3.5.2.1 Transporter

Transport protein selects a specific solute from the extracellular medium. It forms an intermediate complex with a specific solute (the substrate) on the external membrane, which induces the translocation of the substrate to the internal membrane. The binding of the solute to the transport protein enables the transport protein to carry the solute across cell membranes into the internal cell medium, such as in the case of glucose permeation mediated by the glucose transporter protein. Membrane transporters work in concert with foreign compound-metabolizing enzymes to mediate the uptake and efflux of xenobiotics and their metabolites.

Transporter proteins are membrane proteins that control the influx of foreign compounds, including drugs, nutrients, and ions. For example, ATP-binding cassette (ABC-type transporter) and solute carrier (SLC-type transporter) are two major families of membrane transporters for drugs and other foreign compounds. Generally, SLC transporters mediate either influx or efflux of drugs, while ABC transporters mediate unidirectional efflux.

3.5.2.2 Channel

Facilitating membrane permeation of inorganic ions and organic compounds involves channels besides transporters. Channels exist in two primary states: open and closed. In the open state, channels act as pores for selected ions, allowing them to permeate across cell membranes and then return to the closed state. The mechanistic difference between channels and transports results in a marked difference in their turnover rates. The turnover rate constants of typical channels are much larger than those of transporter proteins.

3.5.3 Active Transport

Active transport of a foreign compound across bilayer membranes into the cells is also mediated by membrane transporters. But, unlike facilitated diffusion, active transport is characterized by the requirement of energy as well as the movement of solute against a concentration gradient. Depending on the driving force, active transport can be classified into primary and secondary active transports.

3.5.3.1 Primary Active Transport
Primary active transport is coupled with ATP hydrolysis catalyzed by Na^+ and K^+-ATPases, which provide the energy for the uptake of solute against its concentration gradient. The unidirectional movement of a solute across membranes in mammalian cells is mediated by transporter proteins, such as ATP binding cassette transporters.

3.5.3.2 Secondary Active Transport
In the secondary active transport, the transport of solute across cell membranes uphill against its concentration gradient is coupled with the movement of another solute downhill in accordance with its concentration gradient. Therefore, secondary active transport takes place at the expense of a pre-existing electrochemical gradient of another solute. In this case, the driving force for the uptake of solute across cell membranes is the electrochemical potential stored in a concentration gradient of another solute.

An example of secondary active transport is the uptake of lactose across E. coli membranes, which is coupled with the movement of H^+ downhill in accordance with proton electrochemical potential. Another example is Na^+- Ca^{++} exchange protein, which uses the energy stored in the Na^+ gradient established by Na^+ and K^+-ATPase to export cytosolic Ca^{++}.

3.6 Metabolic Pathways

The physical barriers of cell membranes are attributed to the role of membrane permeability and hydrophobicity in the metabolism of foreign compounds. The capacity of excreting a wide range of lipophilic foreign compounds is the major challenge of xenobiotic metabolic pathways. This challenge is achieved by a combination of low-specificity enzymatic systems in addition to the physical barriers of cell membranes. The low specificity of enzymatic systems makes it possible to metabolize a wide range of lipophilic and hydrophilic compounds.

Foreign compound-metabolizing enzymes are produced based on the information stored within the genes. Xenobiotic metabolic enzymes are present in the liver at much higher concentrations than in other organs. The liver is the primary site for the metabolism of foreign compounds. The liver contains the necessary enzymes for metabolism of drugs and other xenobiotics. Although the liver is the primary site for foreign compound metabolism, virtually all tissue cells have some metabolic activities. Other organs that have significant metabolic activities include the gastrointestinal tract, kidneys, and lungs.

The metabolism of lipophilic foreign compounds involves a set of metabolic pathways. Two major enzyme-catalyzed pathways are: (a) the functionalization in activation reactions, which involves the addition of a functional group to a xenobiotic, and (b) the conjugation or non-conjugation in detoxification reactions, which involves the coupling of an endogenous cofactor molecule to a functionalized foreign compound.

These two pathways are responsible for the metabolism of foreign compounds, but generally they are not implicated in the elimination of endogenous metabolites derived from normal cellular constituents. Metabolic enzymes involved in activation reactions are primarily located in the endoplasmic reticulum of the liver cell, while those involved in detoxification reactions are mainly located in the cytosol.

Drugs are usually lipophilic compounds that can pass plasma membranes and reach the site of action. Before their excretion, drugs often undergo both activation and detoxification reactions. When a drug molecule is oxidized, hydrolyzed, or covalently attached to a hydrophilic species, the molecule becomes more hydrophilic. Thus, drug metabolism is also involved in a process of facilitating its excretion from the body by introducing a hydrophilic functional group to the drug molecule.

3.6.1 Phase I Activation Metabolism

In phase I metabolism, activation enzymes catalyze functionalization reactions by means of oxidation, reduction, or hydrolysis, where a functional group is introduced to the structure of a lipophilic foreign compound. While the addition of a functional polar group results in an increase in the water solubility of the parent compound, the functionalization also activates the foreign compound, which often leads to the formation of a metabolic intermediate or metabolite. Many metabolic intermediates or metabolites are chemically reactive and are ultimately responsible for their toxicities.

3.6.2 Phase II Detoxification Metabolism

In phase II metabolism, a functionalized foreign compound further undergoes conjugation or non-conjugation reaction catalyzed by detoxification enzymes. The conjugation reaction combines a small cofactor molecule with the functionalized foreign compound to form a conjugate. Phase II detoxification metabolic reaction detoxifies metabolic intermediate, rendering it less harmful, and in the meantime largely increases its water solubility, thus facilitating the excretion of foreign compounds from the body.

Figure 3.2 illustrates two major metabolic enzymatic reactions: functionalization and conjugation or non-conjugation reactions for lipophilic foreign compounds. The

(a) Lipophilic foreign compound $\xrightarrow[\text{Activation enzyme}]{\text{Functionalization}}$ Metabolic intermediate or metabolite

(b) Metabolic intermediate + Cofactor ligand $\xrightarrow[\text{Detoxification enzyme}]{\text{Conjugation or non-conjugation}}$ Water soluble conjugate

(c) Overall reactions:

Lipophilic foreign compound $\xrightarrow{\text{Activation enzyme}}$ Metabolic intermediate $\xrightarrow{\text{Detoxification enzyme}}$ Soluble conjugate

Fig. 3.2 Major metabolic pathways for lipophilic foreign compounds

metabolism of hydrophilic foreign compounds can generally bypass phase I functionalization reactions and directly undergo phase II conjugation or non-conjugation reactions.

3.7 Transport to External Cell Compartment

The next process following xenobiotic metabolisms is the transport of conjugates and other metabolites from the internal to the external cell compartment (referred to as phase III metabolism). Preceding their excretion, conjugates require transmembrane movement from the internal to the extracellular cellular compartment. Phase III metabolism of foreign compounds is the step that occurs after metabolic conversion and before their excretion from the body.

Due to membrane lipids acting as the physical barriers, the transport of conjugates or metabolites out of the cells cannot be carried out by free diffusion. Instead, the transmembrane movement of conjugates across cell membranes requires transport proteins such as export pumps. Conjugates may be further processed before being recognized by transport proteins and moving out of the cells. A number of ATP-dependent transport proteins, or export pumps, have been identified in the liver.

ATP-binding transporters are the family of multidrug-resistance proteins. Vectoral transport of solute across cell membranes also plays a major role in the urinary and hepatobiliary excretion of drugs from the blood to the lumen. ABC transporters are able to achieve vectoral transport by extruding lipophilic xenobiotics to the exterior compartment of cells.

3.8 Metabolism Precedes Excretion

Foreign compounds that are excreted from the body include (a) waste products from the digestion of foods, (b) drugs accumulated in the body, (c) chemical substances in the environment, and (d) industrial chemicals in the household. Due to their limited

a. Lipophilic foreign compounds
   Lipophile ⟶ Activation ⟶ Detoxification ⟶ Transport ⟶ Excretion

b. Hydrophilic foreign compounds
   Hydrophile ⟶ Detoxification ⟶ Transport ⟶ Excretion

Fig. 3.3 Foreign Compound Metabolism Precedes Before Excretion

solubility in water, lipophilic compounds require metabolic conversion into hydrophilic metabolites before they are excreted from the body.

Figure 3.3 illustrates metabolic processes that foreign compounds may precede before they are excreted from the body. Water-soluble, hydrophilic foreign compounds do not require metabolic conversion to increase their water solubility. They can proceed directly to the detoxification reaction and transport process before being excreted from the body. However, lipophilic foreign compounds require an activation reaction to increase their water solubility before their excretion.

3.9 Excretion of Foreign Compounds

Excretion is the process by which metabolic waste products are eliminated from an organism. It is primarily carried out by the kidneys, lungs, and skin. The urinary and biliary systems are two primary routes for the excretion of foreign compounds and their metabolites from the body. The excretions of foreign compounds include renal, hepatic, and skin excretions. They are described below.

3.9.1 Renal Excretion

The kidney is the most important organ for the excretion of foreign compounds and their metabolites, including foods, drugs, and other chemicals. More foreign compounds are eliminated from the body by the kidney than by other organs. Renal excretion plays an important role in eliminating foreign compound metabolites resulting from metabolic functionalization reactions. The kidney is efficient in the elimination of toxicants from the body. Kidney excretion is critical to the body's defense against foreign compounds.

Renal excretion involves glomerular filtration, active tubular secretion, and passive tubular re-adsorption. Small compounds are usually able to pass the glomerular filtration barrier, while larger metabolites are excreted via other pathways. After glomerular filtration, a foreign compound or its metabolite may remain in the tubular lumen and be excreted in urine. Specific carriers located on the basolateral membrane of proximal tubule cells can transport foreign compounds from the blood into the epithelial cells for subsequent excretion.

Structurally diverse organic cations and anions, including positively and negatively charged chemicals or drugs and their metabolites, are secreted in the proximal tubule of the kidney. Transporters in the kidney mediate the secretion or reabsorption of many foreign compounds, thereby influencing the plasma levels of their substrates. SLC transporters and ABC transporters are two major classes of secretory transporters in the mammalian kidney.

SLC transporters are involved in moving organic cations across the basolateral membrane. They are also implicated in a variety of organic anions that are secreted in the proximal tubule. Major physiological functions of ABC transporters include the transport of toxic compounds, lipids, and bile salts. ABC transporters are also involved in the secretion of organic cations. They translocate a variety of compounds through cell membranes against the concentration gradients at the expense of ATP hydrolysis.

Besides SLC and ABC transporters, other transporters also participate in the excretion of organic ions. Transporters such as multidrug-resistance-associated proteins localized in the apical brush-border membrane are also responsible for the excretion of conjugated metabolites.

3.9.2 Reabsorption in the Kidney

Transporters in the kidney mediate the secretion or reabsorption of many foreign compounds and thereby influence the plasma levels of their substrates. Kidneys are efficient in the excretion of hydrophilic compounds. Hydrophilic compounds and related ions are readily excreted in the urine, while lipophilic compounds that have high lipid-to-water partition coefficients can be re-adsorbed across the kidney tubules back into the bloodstream.

Tubular re-absorption in the kidney is a major factor contributing to the difficulty in the excretion of lipophilic foreign compounds. Before they are converted into water-soluble polar compounds, lipophilic compounds are not readily eliminated from the body. Therefore, living organisms develop metabolic enzyme systems that are proficient in the conversion of lipophilic into hydrophilic compounds, as discussed above.

3.9.3 Hepatic Excretion

Besides being the main site for the biotransformation of foreign compounds, the liver also plays an important role in removing xenobiotics from the blood after they are absorbed in the gastrointestinal tract. Foreign compounds or their conjugates with moderate molecular weights can be excreted into bile in substantial quantities. Hepatic uptake of organic anions, cations, and bile salts can be carried out by SLC transporters in the basolateral membrane of hepatocytes. SLC transporters mediate either influx or efflux of foreign compounds. Such an uptake process takes place by either facilitated diffusion or secondary active transport.

Moreover, ABC transporters in the canalicular membrane of hepatocytes also mediate the unidirectional efflux of foreign compounds. The excretion of drugs and their metabolites mediated by ABC transporters is carried out uphill against a concentration gradient, where ATP hydrolysis provides the driving force for the transport to occur.

3.9.4 Skin Excretion

The skin consists of an outer layer called the epidermis and an inner layer called the dermis. Its primary functions are temperature control and secretion. Sweat glands in the skin secrete fluid waste as sweat or perspiration. Sweating also maintains the level of salt in the body. Through the production of sweat glands, the skin also plays a role in the excretion of foreign compounds. However, such a role as excretory system for foreign compounds is minimal.

Bibliography

Borst P, Elferink RO (2002) Mammalian ABC transporters in health and disease. Annu Rev Biochem 71:537–592

Burckhardt BC, Burckhardt G (2003) Transport of organic anions across the basolateral membrane of proximal tubule cells. Rev Physiol Biochem Pharmacol 146:95–158

Buxton ILO, Benet LZ (2011) Pharmacokinetics: the dynamics of drug absorption, distribution, metabolism and elimination. In: Brunton LL et al (eds) Goodman & Gilman's the pharmacological basis of therapeutics. McGraw-Hill, New York

Caldwell J, Gardner I, Swales N (1995) An introduction to drug disposition: the basic principles of absorption, distribution, metabolism, and excretion. Toxicol Pathol 23(2):102–114

Chen C-H (2012) Activation and detoxification enzymes: functions and implications. Springer Sciences, New York

Chen C-H (2020) Xenobiotic metabolic enzymes: bioactivation and antioxidant defense. Springer Nature, Cham

Chen LS, Chen C-H (1986) Energetic studies of lactose active transport in Escherichia coli membrane vesicles. Arch Biochem Biophys 246:515–524

Dean M, Rzhetsky A, Allikmets R (2001) The human ATP-binding cassette (ABC) transporter superfamily. Genome Res 11:1156–1166

Dresser MJ, Leabman MK, Giacomini KM (2001) Transporters involved in the elimination of drugs in the kidney: organic anion transporters and organic cation transporters. J Pharm Sci 90:397–421

Giacomini KM, Sugiyama Y (2011) Membrane transporters and drug response. In: Brunton LL et al (eds) Goodman and Gilman's the pharmacological basis of therapeutics. McGraw-Hill, New York

Hediger MA, Romero MF, Peng JB et al (2004) The ABCs of solute carriers: physiological, pathological and therapeutic implications of human membrane transport proteins. Pflugers Arch 447:465–468

Koepsell H (1998) Organic cation transporters in intestine, kidney, liver, and brain. Annu Rev Physiol 60:243–266

Lehman-McKeeman LD (2010) Absorption, distribution, and excretion of toxicants. In: Chapter 5 in Casarett & Doull's Essentials of Toxicology. McGraw-Hill, New York

Metabolic Conversion of Foreign Compounds

During biotransformation involving activation and detoxication processes, foreign compounds are metabolized to form more polar products so as to facilitate their elimination from the body. Particularly, the conversion of lipophilic, non-polar species to polar, water-soluble, hydrophilic compounds is essential to facilitate their excretion from the body through urea or bile. Metabolic processes include phase I metabolism engaging in functionalization reactions, phase II metabolism involving conjugation reactions, and phase III metabolism eliminating metabolites from the body.

In phase I functionalization, a functional polar group, such as -OH or -COOH, is generated on a foreign compound to form an intermediate through oxidation, reduction, or hydrolysis reaction. Functionalization increases the water solubility of the parent compound and paves the way for the next phase of metabolism.

In phase II conjugation reaction, the functionalized compound is combined with an endogenous substrate, such as glucuronic acid, glutathione, or sulfate, to produce a conjugate. The conjugation reaction greatly increases the water solubility of a foreign compound, thus facilitating its excretion from the body. Whereas, a foreign compound that is polar and water soluble can bypass the functionalization reaction and directly involve the phase II conjugation reaction.

The major enzyme system that deals with chemicals and drugs in activation metabolism is cytochrome P450s. Other enzyme systems include dehydrogenases, oxidases, esterases, and reductases. A number of conjugating enzyme systems include glucuronosyltransferases, sulfotransferases, glutathione S-transferases, etc. Besides metabolic enzymes, environmental and genetic factors can cause differences in foreign compound metabolism.

4.1 Phase I Activation Metabolism

Oxidation, reduction, or hydrolysis reactions catalyzed by phase I activation enzyme result in the introduction of a functional group, which leads to the modification of a foreign compound and increase in its water solubility. In the case of a drug, the introduction of a functional group may lead to an alteration in its biological properties. The product of phase I metabolism subsequently serves as the substrate for the phase II conjugation reaction.

Major reactions catalyzed by phase I activation enzymes in metabolic pathways include N-dealkylation, O-dealkylation, aliphatic and aromatic hydroxylation, N-oxidation, S-oxidation, epoxidation, and hydrolysis. These phase I reactions are briefly described below, where the symbol Φ or Φ' denotes an aromatic side group, while R or R' represents a straight or cyclic aliphatic side group.

4.1.1 Oxidation Reactions

Oxidation reactions of phase I activation enzymes include N-oxidation and S-oxidation.

4.1.1.1 N-oxidation
N-oxidation results in forming hydroxylamine, as presented in the following chemical equation for the metabolism of chlorpromazine or the conversion of nicotine.

$$R\text{-}NH_2 \rightarrow R\text{-}\underset{|}{\overset{H}{N}}\text{-}OH$$

4.1.1.2 S-oxidation
S-oxidation results in producing oxysulfide, as presented in the following chemical equation for the oxidation of chlorpromazine.

$$R\text{-}\underset{|}{\overset{R}{S}}H_2 \rightarrow R\text{-}\underset{|}{\overset{R}{S}}=O$$

4.1.2 Hydroxylation Reactions

Hydroxylation reactions include aromatic hydroxylation and aliphatic hydroxylation.

4.1.2.1 Aromatic Hydroxylation

Aromatic hydroxylation reaction applies to a foreign compound that contains an aromatic ring, as presented in the following chemical equation for aromatic hydroxylation leading to the generation of phenolic products.

$$R - \Phi \rightarrow R - \Phi - OH$$

4.1.2.2 Aliphatic Hydroxylation

Aliphatic hydroxylation applies to a foreign compound that contains an aliphatic chain, as presented in the following chemical equation for the pathway for many drugs.

$$R - \Phi - CH_3 \rightarrow R - \Phi - CH_2 - OH$$

4.1.3 Dealkylation Reactions

Dealkylation reactions include O-dealkylation and N-dealkylation.

4.1.3.1 O-dealkylation

O-dealkylation results in the loss of the alkyl group attached to oxygen, as presented in the following chemical equation for the conversion of ethoxycoumarin to hydroxycoumarin.

$$R - \Phi - OCH_3 \rightarrow R - \Phi - OH + CH_2O$$

4.1.3.2 N-dealkylation

N-dealkylation results in the loss of alkyl group attached to nitrogen, as presented in the following chemical equation for N-dealkylation of benzphetamine, where the removal of an *N*-alkyl group forms an amine.

$$\Phi\text{-}\underset{|}{\overset{H}{N}}\text{-}CH_3 \rightarrow \Phi\text{-}\underset{|}{\overset{H}{N}}\text{-}H + CH_2O$$

4.1.4 Hydrolysis

Hydrolysis is a pathway for a number of esters or amides, such as esterase or amidase- catalyzed reactions, as presented in the following chemical equation for the hydrolysis of oleyl oleate or the conversion of propanil to dicholoroaniline.

$$\Phi - NH - CO - CH_2CH_3 \rightarrow \Phi - NH_2 + CH_3CH_2COOH$$

Fig. 4.1 Functional groups introduced by phase I metabolic reactions

Phase I reaction	Functional group
N-oxidation	N–OH
S-oxidation	S=O
Aromatic hydroxylation	Φ–OH
Aliphatic hydroxylation	R–OH
O-dealkylation	R–OH
N-dealkylation	–NH$_2$
Hydrolysis	–NH$_2$ or –COOH
Epoxidation	–C–C– \\ / O

4.1.5 Epoxidation

Epoxidation is a pathway in metabolizing many carcinogenic compounds, as presented in the following chemical equation for the epoxidation of diadinoxanthin or the conversion of aflatoxin to aflatoxin-epoxide.

$$-\Phi - \underset{H}{RC} = CR^* \quad \rightarrow \quad -\Phi - RC \underset{O}{\overset{\diagdown \; \diagup}{-}} CHR^*$$

Figure 4.1 summarizes the functional groups that are introduced by phase I metabolic reactions. Phase I metabolic reaction leads to the introduction of a functional group (e.g., –OH, –COOH, –SH, –O– or NH$_2$) to a lipophilic foreign compound, The results modify the chemical structure of the parent compound to form the metabolite. Such a modified compound then serves as the substrate for phase II conjugation or non-conjugation enzymes.

4.2 Phase II Detoxification Metabolism

Phase II detoxification metabolism involves conjugation and non-conjugation reactions.

In conjugation reactions, an ionized group is attached to a foreign compound to make it more soluble in water. Conjugation reactions include glucuronide, glutathione, sulfonate, amino acid, N-acetyl, and methyl conjugations. Non-conjugation reactions include quinone reductase and epoxide hydrolase catalytic reactions. These reactions are described below:

4.2.1 Conjugation Reactions

Metabolic pathways in phase II detoxification metabolism include glucuronide, glutathione, sulfate, and amino acid conjugation reactions, which are catalyzed by respective phase II transferase enzymes. A conjugate is formed in the combination of the substrate (the phase I metabolic product) with a small endogenous molecule.

Conjugation reactions usually require the substrate to have oxygen, nitrogen, or sulfur atoms to serve as an acceptor site for the hydrophilic moiety, such as glutathione, glucuronic acid, sulfonate, or acetyl group. Such phase II enzymatic reactions inactivate potentially toxic metabolites generated in phase I activation reactions, where the water solubility of foreign compounds is greatly improved so as to facilitate their elimination.

Major conjugation reactions catalyzed by phase II detoxification enzymes in metabolic pathways include glucuronide, sulfate, glutathione, acetyl, methyl, and amino conjugations. Typical phase II conjugation and non-conjugation reactions are briefly described below. The symbol Φ or Φ' denotes an aromatic side group, and R or R' represents a straight or cyclic aliphatic side group.

4.2.1.1 Glucuronide Conjugation

The glucuronide conjugation reaction transfers glucuronic acid from the cofactor UDP-glucuronic acid to a substrate (functionalized foreign compound) to form the glucuronide metabolite. The functional group that is ready for glucuronide conjugation includes phenol and aromatic amines, as presented in the following chemical equation for the conversion of hydroxydiazepam to hydroxydiazepam glucuronide.

$$-\Phi-OH \quad + \quad \text{UDP-glucuronic acid} \quad \rightarrow \quad -\Phi-O-\text{glucuronide}$$

The formation of glucuronide conjugates is a metabolic pathway for drugs that contain a carboxylic acid group. The formation of acyl glucuronides results in an increase in aqueous solubility as compared to the parent drug, thus facilitating their excretions in urine or bile.

4.2.1.2 Glutathione Conjugation

Glutathione is a tripeptide (glutamate-cysteine-glycine) that contains a sulfhydryl group. Glutathione conjugation transfers glutathione to reactive electrophiles, where GSH is the reduced form of glutathione as presented in the following chemical equation for the conversion of paracetamol to glutathione derivative.

$$HO-\Phi-OH \quad + \quad GSH \quad \rightarrow \quad HO-\Phi-GSH$$

Glutathione plays a role in detoxification to avoid foreign compounds reacting with cellular targets. Glutathione usually combines with their metabolites or anticancer drugs to form water-soluble glutathione S-conjugates, so as to be ready for excretion from the body.

4.2.1.3 Sulfonate Conjugation

Sulfonate conjugation transfers sulfonate from 3'-phosphoadenosine-5'-phosphosulfate (PAPS) to the –OH group of aromatic or aliphatic compound. Sulfonate conjugation is an important pathway in the metabolism of amino group, as presented in the following chemical equation for the conversion of hydroxycoumarin to sulfate derivative.

$$-\Phi-OH \;+\; PAPS \;\rightarrow\; -\Phi-O-SO_2-OH \;+\; PAP$$

4.2.1.4 Amino Acid Conjugation
Amino acid conjugation reaction is important in the metabolism of an organic acid. The carboxylic group of an organic acid is conjugated with an amino acid, such as glycine. Glycine is the most common amino acid in amino acid conjugation reaction. Amino acid conjugation reaction increases the aqueous solubility of bile acids, which are converted to glycine and taurine N-acyl amidates, as presented in the following chemical equation:

$$-\Phi-COOH \;+\; H_2NCH_2COOH \;\rightarrow\; -\Phi-CONHCH_2COOH$$

4.2.1.5 N-acetyl Conjugation
N-acetyl conjugation is important in the metabolism of environmental chemicals and drugs that contain an aromatic amine or hydrazine group, as presented in the following chemical equation for the conversion of $-NH_2$ on dapsone to $-NH$-acetyl.

$$-\Phi-NH_2 \;+\; \text{acetyl-CoA} \;\rightarrow\; -\Phi-NH-\text{acetyl} \;+\; CoA$$

N-acetyl conjugation transfers acetyl from the cofactor acetyl-CoA to $-NH_2$ group of the metabolite. N-acetyl conjugate is often less water soluble than the parent compound.

4.2.1.6 Methyl Conjugation
Foreign compounds can undergo O-, N-, or S- methylation using S-adenosyl-methionine (AdoMet) as the methyl donor, as presented in the following chemical equation:

$$-\Phi-R-NH_2 \;+\; AdoMet \;\rightarrow\; -\Phi-R-NH-CH_3$$

such as the conversion of $-NH_2$ group on dopamine to $-NH$-CH_3.

Methyl conjugate has less water solubility than the parent compound. Methyl conjugation is an important pathway in metabolism of drugs. The activities of three human drug-metabolizing methyltransferase enzymes are catechol-O-methyltransferase, thiopurine methyltransferase, and thiol methyltransferase. Table 4.1 presents typical cofactors and conjugation groups in phase II metabolic reactions.

4.2 Phase II Detoxification Metabolism

Table 4.1 Cofactors and conjugation groups in phase II reactions

Phase II reaction	Cofactor	Conjugation group
Glucuronidation	UDP-glucuronic acid	Glucuronic acid
Sulfonation	3′-phosphoadenosine-5′-phosphosulfate	Sulfonate
Glutathione conjugation	Glutathione	Glutathione
Acetylation	Acetyl-Coenzyme A	Acetyl group
Methylation	S-adenosyl-methionine	Methyl group
Amino acid conjugation	Glycine	Amino acid

4.2.2 Non-conjugation Reactions

Non-conjugation reactions that are involved in detoxification metabolism include quinone reductase and epoxide hydrolase catalytic reactions. They are briefly described below:

4.2.2.1 Quinone Reductase Catalytic Reactions

Quinones can be reduced to hydroquinones by NADPH-quinone oxidoreductase, as presented in the following chemical equation for the conversion of menadione to hydroquinone.

$$R - \Phi = O \quad \rightarrow \quad R - \Phi - OH$$

Studies have been carried out on the oxidation of hydroquinone to quinone via semiquinone, which results in superoxide generation.

Superoxide generation in biological systems is achieved through complex enzymatic systems such as xanthine oxidase. The xanthine and xanthine oxidase-catalyzed reduction of estrogen quinone has been studied to elucidate the role of metabolic redox cycling in estrogen metabolism.

4.2.2.2 Epoxide Hydrolase Catalytic Reactions

Many epoxides are intermediary metabolites that are formed during the oxidation of unsaturated aromatic or aliphatic compounds, as presented in the following chemical equation for the conversion of styrene 7,8-epoxide to styrene 7,8-glycol.

$$\Phi - HC - CH_2 \quad \rightarrow \quad \Phi - HC - CH_2$$
$$\begin{array}{c} \backslash \ / \\ O \end{array} \qquad\qquad \begin{array}{cc} | & | \\ HO & OH \end{array}$$

Epoxide hydrolases play an important role in the detoxification of electrophilic epoxides. Biocatalytic reactions involving epoxides include epoxide hydrolases. Epoxide hydrolases are important industrial biocatalysts that catalyze the hydrolysis of epoxides to the corresponding diols.

4.3 Toxification and Detoxification

As described above, the metabolism of lipophilic foreign compounds is composed of two processes: activation and deactivation. Activation is referred to as the process by which foreign compounds are transformed to generate intermediates or metabolites. If the generated intermediates are highly active, some of them can interact with cellular components before the phase II detoxification reaction takes place. Metabolic conversion of reactive intermediates to metabolites that are no longer capable of causing cellular damage is referred to as deactivation.

Phase II reactions generally yield stable, non-reactive metabolites that are available for excretion from the body.

4.3.1 Toxification Activation

The overall metabolism of lipophilic foreign compounds is a detoxification process. A metabolic chemical intermediate is potentially reactive and more toxic than the parent compound. Generally, reactive intermediates generated through functionalization reactions mediated by CYP450 and other phase I enzymes are most implicated in phase I-mediated activation reactions. Reactive chemical intermediates are capable of interacting with cellular components, such as proteins, nucleic acids, and lipids, and are ultimately responsible for the toxicity of foreign compounds.

A typical example is the transformation of 4-ipomeanol by the cytochrome 4B1 enzyme into an active intermediate, α,β-unsaturated dialdehyde, which exhibits toxicity and leads to 4-ipomeanol-induced lung injuries. Highly active intermediates generated by foreign compound metabolism also include epoxides, radicals, and carbonium ions. Moreover, chemically active intermediates can further induce toxicity or carcinogenicity by interacting with molecular oxygen to yield reactive oxygen species, such as hydrogen peroxide and hydroxyl (OH^-) and superoxyl anion ($O_2^{\cdot}$) radicals. Reactive oxygen species and free radicals are believed to be involved in degenerative diseases and cancers.

4.3.2 Deactivation of Toxicity

The body's major defensive mechanisms against foreign compounds or their reactive metabolites include enzymatic reactions and non-enzymatic reactions. Phase II conjugation reaction carries out enzymatic defense by combining a functionalized foreign compound with an enzyme cofactor, thus facilitating the excretion of xenobiotic from the body. While, non-enzymatic reaction utilizes endogenous compounds, such as antioxidant glutathione, to prevent damage to cellular components caused by reactive oxygen species, including peroxides and free radicals.

While the vast majority of phase II enzyme-catalyzed metabolizing reactions result in the deactivation of foreign compounds, in minor cases, phase II enzyme may be implicated in the toxicity of foreign compounds. A typical example of phase II enzyme-mediated toxicity is the glutathione conjugation of epoxide, which exhibits toxic effects in the kidneys.

Besides phase II enzyme-catalyzed metabolism, there are other defense mechanisms against chemically active intermediates and reactive oxygen species. These include antioxidant enzymes, such as glutathione peroxidase and superoxide dismutase, as well as other antioxidant molecules, such as uric acid and vitamin E.

4.3.3 Activation Versus Deactivation: Competing Pathways

The amount of chemically active intermediate present in a body largely depends on the competing pathways between activation and deactivation metabolisms. The body requires a delicate balance between the rates of activation and deactivation. Any factors that interrupt this delicate balance can affect the implications of foreign compounds.

When phase I enzymes are over expressed, activation pathways will overwhelm deactivation pathways. As activation pathways assume a greater role, an over production of reactive intermediates happens. As a result, the accumulation of reactive intermediates in the body occurs, which can lead to a potential increase in the interaction of foreign compounds with cellular components.

A low expression of conjugating enzyme activity can lead to a low rate of conjugation reactions. An insufficiency of cofactors, such as glutathione, sulfate, or glucuronic acid, can also result in a deficiency in conjugation pathways, due to the rate of utilization exceeding the supply of a cofactor. Besides, the environmental factors can also attribute activation pathways to deactivation pathways under the following two conditions: (a) when enzyme systems that catalyze activation pathways are selectively induced; and (b) when enzyme systems that catalyze deactivation pathways are saturated due to the exposure of a large quantity of chemicals.

Since foreign compounds rely heavily on phase II enzymes for the detoxification, an inactive or deficient phase II conjugation pathway may lead to an increased vulnerability to the toxicity of foreign compounds. Activation pathways may become more dominant than deactivation pathways under the circumstances, such as the effectiveness of conjugation reactions, the association with environmental factors, and the metabolic enzyme polymorphisms.

Extensive studies have showed that metabolic activation and detoxification enzymes exhibit genetic polymorphisms, with two or more variants differing slightly in their amino acid sequences. Such enzyme polymorphisms can give rise to the differences in the ability of individuals to metabolize foreign compounds.

Bibliography

Bucko M, Kaniakova K, Hronska H et al (2023) Epoxide hydrolases: multipotential biocatalysts. Int J Mol Sci 24(8):7334

Camilleri P, Buch A, Soldo B, Hutt AJ (2018) The influence of physicochemical properties on the reactivity and stability of acyl glucuronides. Xenobiotica 48(9):958–972

Cavalieri EL, Rogan EG (2005) The approach to understanding aromatic hydrocarbon carcinogenesis. The central role of radical cations in metabolic activation. Pharmacol Ther 55:183–199

Chakraborty C, Davis DL, Dey SK (1990) Estradiol-15 alpha-hydroxylation: a new avenue of estrogen metabolism in peri-implantation pig blastocysts. J Steroid Biochem 35(2):209–218

Chen C-H (2012) Activation and detoxification enzymes: functions and implications. Springer, New York

Chen C-H (2020) Xenobiotic metabolic enzymes: bioactivation and antioxidant defense. Spring Nature, Cham

Cribb AE, Peyrou M, Muruganandan S et al (2005) The endoplasmic reticulum in xenobiotic toxicity. Drug Metab Rev 37:405–442

de Vries EJ, Janssen DB (2003) Biocatalytic conversion of epoxides. Curr Opin Biotechnol 14(4): 414–420

Goss R, Latowski D, Grzyb J et al (2007) Lipid dependence of diadinoxanthin solubilization and de-epoxidation in artificial membrane systems resembling the lipid composition of the natural thylakoid membrane. Biochim Biophys Acta 1768(1):67–75

Hayes JD, Flanagan JU, Jowsey IR (2005) Glutathione transferases. Annu Rev Pharmacol Toxicol 45:51–88

Holmberg AA, Weidolf L, Necander S et al (2022) Characterization of clinical absorption, distribution, metabolism, and excretion and pharmacokinetics of velsecorat using an intravenous microtracer combined with an inhaled dose in healthy subjects. Drug Metab Dispos 50(2): 150–157

Maynard MS, Brumback D, Itterly W et al (1999) Metabolism of [(1)(4)C]prometryn in rats. J Agric Food Chem 47(9):3858–3865

Meisel P (2002) Arylamine N-acetyltransferases and drug response. Pharmacogenomics 3:349–366

Meyer UA (1996) Overview of enzymes of drug metabolism. J Pharmacokinet Biopharm 24(5): 449–459

Najmi AA, Bischoff R, Permentier HP (2022) *N*-Dealkylation of Amines. Molecules 27(10):3293

Negishi M, Pedersen LG, Petrotchenko E et al (2001) Structure and function of sulfotransferases. Arch Biochem Biophys 390:149–157

Ohmiya Y, Mehendale HM (1980) Uptake and metabolism of chlorpromazine by rat and rabbit lungs. Drug Metab Dispos 8(5):313–318

Park BK, Kitteringham NR, Maggs JL et al (2005) The role of metabolic activation in drug-induced hepatotoxicity. Annu Rev Pharmacol Toxicol 45:177–202

Potęga A (2022) Glutathione-mediated conjugation of anticancer drugs: an overview of reaction mechanisms and biological significance for drug detoxification and bioactivation. Molecules 27(16):5252

Pumford NR, Halmes NC (1997) Protein targets of xenobiotic reactive intermediates. Annu Rev Pharmacol Toxicol 37:91–117

Rostrup-Nielsen T, Pedersen LS, Villadsen J (1990) Thermodynamics and kinetics of lipase catalysed hydrolysis of oleyl oleate. J Chem Technol Biotechnol 48(4):467–482

Roy D, Kalyanaraman B, Liehr JG (1991) Xanthine oxidase-catalyzed reduction of estrogen quinones to semiquinones and hydroquinones. Biochem Pharmacol 42(8):1627–1631

Shonsey EM, Wheeler J, Johnson M et al (2005) Synthesis of bile acid coenzyme a thioesters in the amino acid conjugation of bile acids. Methods Enzymol 400:360–373

Singh SK, Husain SM (2018) A redox-based superoxide generation system using quinone/quinone reductase. Chembiochem 19(15):1657–1663

Takahashi RH, Ma S, Yue Q et al (2016) Absorption, metabolism and excretion of cobimetinib, an oral MEK inhibitor, in rats and dogs. Xenobiotica 47(1):50–65

Tukey RH, Strassburg CP (2000) Human UDP-glucuronosyltransferases: metabolism, expression, and disease. Annu Rev Pharmacol Toxicol 40:581–616

Weinshilboum RM, Otterness DM, Szumlanski CL (1999) Methylation pharmacogenetics: catechol O-methyltransferase, thiopurine methyltransferase, and histamine N-methyltransferase. Annu Rev Pharmacol Toxicol 39:19–52

Phase I Activation Enzymes

5

A vast majority of chemical reactions in living cells exhibit kinetic barriers that prevent the reactions from occurring spontaneously. Kinetic barriers can be overcome by a variety of enzymes that act as catalysts, making the reactions energetically favorable. Consequently, the reactions can occur at the rates required for maintaining cell functions.

Enzymes make up the largest and most highly specialized class of protein molecules. They are highly effective catalysts for a large diversity of chemical reactions. The substrate is a small chemical molecule on which the enzyme exerts its catalytic action. On the basis of the nature of chemical reactions that enzymes catalyze, enzymes can be divided into the following six major classes: (a) oxidoreductases that catalyze the oxidation and reduction reactions, (b) transferases that catalyze the transfer of functional groups, (c) hydrolases that catalyze the hydrolysis reactions, (d) lyases that catalyze the reactions involving the removal or addition of a group to double bonds, (e) isomerases that catalyze the reactions involving the intramolecular rearrangement, and (f) ligases that catalyze the reactions that join together two molecules.

Foreign compound-metabolic phase I enzymes and phase II enzymes belong to three of the above six classes of enzymes: oxidoreductases, hydrolases, and transferases. Foreign compound-metabolizing enzymes are primarily present in the endoplasmic reticulum of the cells and in the cytosol. Being produced from the information stored within the genes, they are present in most tissues, with the highest levels located in the liver and intestines.

In the metabolism of foreign compounds, phase I enzymes catalyze functionalization reactions and are hence referred to as activation enzymes or activators. Phase I enzymes are composed of oxidases, hydrolases, and reductases. Oxidases are such as CYP450, flavin-containing monooxygenases, amine oxidases, lipoxygenases, aldehyde and xanthine oxidases, alcohol dehydrogenases, and peroxidases, while reductases include nitroreductases and azoreductases. Hydrolases include carboxylesterases and epoxide hydrolases.

5.1 Activation of Foreign Compounds

A large majority of foreign compounds are lipophilic compounds that are capable of passing through cell membranes and entering cells. They are then transported by lipoproteins in body fluids. The ultimate goal of xenobiotic-metabolic enzymes is to convert lipophilic foreign compounds into water-soluble species so as to facilitate their elimination from the body.

Phase I enzymes catalyze functionalization reactions by the introduction of functional groups to foreign compounds by means of the oxidation, hydrolysis, or reduction reactions catalyzed by oxidoreductases or hydrolases. In many cases, the addition of a functional group results in the formation of an intermediate that is more chemically active and more toxic than the parent compound. Such highly reactive electrophilic metabolites are able to react with cellular molecules, such as proteins, nucleic acids, or lipids, causing toxicity to cells and organs. Phase I enzymes are therefore referred to as activation enzymes or activators.

5.2 Activation Enzymes

Oxidases, reductases, and hydrolases are three major groups of Phase I activation enzymes. Oxidases and reductases catalyze oxidation and reduction reactions, respectively, while hydrolases catalyze hydrolysis reactions with the introduction of water. A large majority of organic chemicals undergo oxidation reactions during activation metabolism. In comparison, enzymatic reduction reactions are less investigated as compared to oxidation and hydrolysis reactions.

For example, benzene rings such as biphenyl commonly undergo aromatic hydroxylation reaction, leading to the formation of phenolic products. Other compounds that contain nitrogen may be oxidized to form nitrooxides or hydroxylamines. Many drugs are metabolized through aliphatic or aromatic hydroxylation.

Major enzymes involved in the metabolic activation of foreign compounds are summarized in Table 5.1.

Table 5.1 Major activation enzymes in foreign compound metabolism

Oxidative enzymes	Reductive enzymes	Hydrolytic enzymes
Cytochrome P450	Nitroreductase	Carboxylesterase
Flavin-monooxygenase	Azoreductase	Epoxide hydrolase
Amine oxidase		
Lipoxygenase		
Alcohol dehydrogenase		
Aldehyde oxidase		
Xanthine oxidase		
Peroxidase		

5.2 Activation Enzymes

5.2.1 Oxidative Enzymes

Table 5.1 shows that cytochrome P450, flavin monooxygenase, amine oxidases, lipoxygenases, alcohol dehydrogenase, aldehyde and xanthine oxidases, and peroxidase are classified as oxidative enzymes (oxidases). While, alcohol dehydrogenase and aldehyde oxidase can catalyze both oxidative and reductive reactions, depending on the substrate or conditions. The functional properties of these oxidative enzymes are discussed as follows:

(a) Cytochrome P450

Among foreign compound-metabolic enzymes, cytochrome P450 (CYP450) is the most actively studied phase I enzyme. CYP450 refers to a unique family of heme proteins. It is responsible for the phase I metabolism of approximately 75% of known pharmaceuticals.

The term P450 is designated because its reduced form binds carbon monoxide to produce a maximum optical absorption around 450 nm.

CYP450s are the most important enzyme family in phase I metabolism because they are responsible for metabolizing the vast majority of therapeutic drugs and other foreign compounds. The family of CYP450 is complex and diverse in catalytic activities. There are over 50 CYP450 isozymes that have been identified in humans. CYP450 contains a heme that is bound to the polypeptide chain. The heme contains one atom of iron. CYP450s utilize O_2 and H^+ (from NADPH) to carry out the oxidation of foreign compounds.

The function of CYP450 is to be a monooxygenase that catalyzes the insertion of one atom of oxygen molecule into the substrate. A molecule of oxygen first binds to the heme moiety and is subsequently cleaved. One oxygen atom forms by chemical binding with the substrate, while the remaining oxygen atom is reduced to yield water. CYP450s catalyze the metabolism of a large number of structurally diverse chemicals, including N-and O-dealkylation, aliphatic and aromatic hydroxylation, N-and S-oxidation, deamination, cleavage of esters, and epoxidation of a double bond.

A diversity of foreign compounds that are catalyzed by CYP450 include food additives, drugs, chemicals and environmental substances. In addition, CYP450s also catalyze the reactions involving endogenous compounds, such as cholesterol, steroid hormones, and fatty acids. CYP450s also play a key role in the adverse effects of foreign compounds since they catalyze the conversion of foreign compounds to reactive intermediates and the formation of electrophilic metabolites.

CYP450 isozymes are also very important for the pharmaceutical industry and have far-reaching implications in medicine, especially in the activation of therapeutic agents.

Microsomal CYP450 isozymes involved in metabolizing foreign compounds are responsible for the breakdown of medications, which occurs mostly in the liver.

There are also significant variations in the levels of enzyme expression present in CYP450 isozymes among individuals due to the presence of genetic polymorphisms and differences in gene regulation. Genetic variants in CYP450 have a potentially functional impact on the efficacy and adverse effects of drugs. Besides, the superfamily of CYP450s consists of over 50 functional genes in the 1, 2, and 3 families. Different ethnic groups may also exhibit variations in the distribution of these genes.

Liver microsomes contain numerous CYP450 isozymes, such as CYP1A2, CYP1A2, and CYP2A6. Each of these enzymes has the potential to catalyze various types of reactions. Besides the liver, CYP450s are also present in other organs, such as the intestine, lung, and kidney. Environmental and genetic factors can play a role in the expression of CYP450 activities. Besides, CYP450 expression in its catalytic reactions can differ markedly as a result of exposure to dietary and environmental enzyme inducers.

(b) Prostaglandin H Synthase

Prostaglandin H synthase (PGHS) is known as cyclooxygenase. It is the initial enzyme in arachidonate metabolism, resulting in the formation of prostanoids such as prostaglandins. PGHS enzymes also provide interesting aspects for pharmacology and toxicology, including the oxidation of chemicals and the altered synthesis of prostanoids. PGHS catalyzes the oxidation of arachidonic acid to prostaglandin in reactions utilizing the activities of cyclooxygenase and peroxidase.

PGHS can also bioactivate many chemical carcinogens to their ultimate mutagenic and carcinogenic forms. PGHS-dependent bioactivation is most important in extrahepatic tissues with low monooxygenase activity. The oxidation of foreign compounds by PGHS often forms metabolites or adducts to cellular macromolecules, which are specific for peroxidase- or peroxyl radical-dependent reactions. PGHS also plays a role in the bioactivation of several polycyclic aromatic hydrocarbons and aromatic amines.

(c) Flavin-Containing Monooxygenases

Flavoproteins are flavin-dependent enzymes that consist of monooxygenases, oxidases, and dehydrogenases. Flavin-containing monooxygenases (FMO) are another superfamily of phase I enzymes involved in the metabolism of foreign compounds. FMOs are expressed at high levels in the liver and are bound to the endoplasmic reticulum.

The family of FMO consists of a group of enzymes that catalyze chemical reactions through the binding of cofactor flavin. FMOs are oxidative enzymes that catalyze the oxygenation of nitrogen-, sulfur-, phosphorus-, and other nucleophilic heteroatom-containing chemicals. FMOs were found to be associated with the detoxification of nucleophilic heteroatom-containing foreign compounds, such as chemicals, drugs, and food components.

The reaction catalyzed by FMO involves one-step two-electron substrate oxygenation, as opposed to two sequential one-electron oxidations as catalyzed by CYP450. FMO-catalyzed reactions usually convert foreign compounds into relatively polar metabolites. Kinetics studies revealed that the mechanisms of FMO1,

5.2 Activation Enzymes

2, 3, and 4 isozymes are similar but differ in the substrate specificities. Among FMO isozymes, FMO3 is most associated with FMO-mediated foreign compound metabolism. FMO3 allelic variation could contribute to the difference in FMO3-dependent metabolism of chemicals among individuals.

Among limited information available for FMO-mediated activation, human FMO oxygenates nucleophilic heteroatom-containing chemicals. Such oxidation reactions generally convert foreign compounds into harmless, polar, readily excreted metabolites. However, FMO sometimes activates foreign compounds into metabolic intermediates that can cause toxic effects.

(d) Amine Oxidases

Amine oxidases are widely distributed throughout the body, in the liver, intestines, and other organs. They are involved in the oxidative deamination of primary, secondary, and tertiary amines. Among amino oxidases, monoamine oxidases (MAOs) are most extensively studied in terms of their involvement in foreign compound metabolism.

MAOs contain the covalently bound cofactor FAD. They are classified as flavoproteins. MAOs are composed of two structurally related flavin-containing enzymes. Mitochondrial outer membrane-bound flavoproteins catalyze the oxidative deamination of amines. Oxygen is used to remove an amine group in MAO-catalyzed oxidation of aliphatic and aromatic amines.

The products of MAO-catalyzed reactions are ammonia, hydrogen peroxide and aldehyde, which are potentially toxic.

The produced aldehyde may be further metabolized by aldehyde dehydrogenase or aldehyde oxidase to form the corresponding carboxylic acid. Hepatic monoamine oxidases have a key defense role in the detoxification of amines. Foreign compounds that are metabolized by monoamine oxidases include tyramine in foods and beverages, 2-phenylethylamine in dietary sources, benzylamine in mouth washes, and 2-phenylpropanolamine in decongestants and cough medicines.

(e) Lipoxygenases

Lipoxygenases are major enzymes in the oxidation of foreign compounds. They are a family of non-heme iron-containing enzymes that catalyze the oxygenation of fatty acids to corresponding lipid hydroperoxide. Lipoxygenase-catalyzed reactions involve a fatty acid substrate with two cis double bonds separated by a methylene group.

Lipid oxidation is an important reaction in physiological and toxicological processes. Lipoxygenases mediate the oxidative metabolism of foreign compounds, including industrial chemicals, pesticides, and drugs. Lipoxygenases are present in many mammalian tissues, such as liver, lung, kidney, and colon. Unlike CYP450, which inserts one oxygen atom into the substrate, lipoxygenases oxidize arachidonic acid, an essential polyunsaturated fatty acid, by inserting two oxygen atoms.

Lipoxygenases also catalyze N-dealkylation and epoxidation. Examples of lipoxygenase-mediated N-demethylation of drugs include aminopyrine and chlorpromazine. Lipoxygenases are also important in the final step of epoxidation of polycyclic aromatic hydrocarbons such as benzo[a]pyrene.

(f) Alcohol Dehydrogenase

Ethanol is metabolized to acetaldehyde by alcohol dehydrogenase (ADH). The formation and degradation of acetaldehyde depend on the activities of alcohol and aldehyde dehydrogenases.

ADHs are oxidative enzymes that break down alcohols. They are probably the most important dehydrogenases involved in the metabolism of alcohols. ADHs catalyze the conversion of primary and secondary alcohols to aldehydes or ketones, respectively.

The ADH conversion reaction involves the reduction of the coenzyme nicotinamide dinucleotide NAD^+ to NADH. ADH-catalyzed reaction is reversible. The produced carbonyl compound can be reduced to alcohol. The conversion product, aldehyde, is usually toxic. Hence, further oxidation of aldehyde to acid is a vital detoxification reaction.

(g) Aldehyde Oxidase

Aldehyde oxidase is important in drug oxidation, activation, and detoxification. The enzyme exhibits oxidase activity toward various aliphatic and aromatic aldehydes. High levels of aldehyde oxidase activity are present in the liver. In aldehyde oxidase, the ultimate source of oxygen inserted into a substrate is water, not O_2. While aldehyde oxidase catalyzes the oxidation of foreign compounds, the enzyme is also involved in the reduction reactions, such as reduction of nitroaromatics to hydroxylamines.

(h) Xanthine Oxidase

Similar to aldehyde oxidase, xanthine oxidase is a molybdoenzyme, but differs in substrate specificities. Xanthine oxidase is also important in drug oxidation, activation, and detoxification. Many features of aldehyde oxidase apply to xanthine oxidase. Xanthine oxidase also exhibits oxidase activity toward various aliphatic and aromatic aldehydes. High levels of xanthine oxidase activity are found in other tissues besides the liver.

(i) Peroxidases

The family of peroxidases includes several peroxidases in addition to prostaglandin H synthase. Prostaglandin H synthase is a dual-function enzyme consisting of a peroxidase and a cyclooxygenase. It is one of the most extensively studied peroxidases that are involved in the activation of foreign compounds, in particular in tissues with low CYP450 activity. In addition to peroxide oxidation, prostaglandin H synthase also catalyzes a number of diverse oxidation reactions involving phenols and aromatic amines.

5.2.2 Reductive Enzymes

Compared to enzymatic oxidation, hydrolysis, and conjugation reactions, there is less research information available for enzymatic reduction reactions on foreign compound metabolism. Nitroreductases and azoreductases are two known phase I reduction enzymes (reductases). These enzymes catalyze reduction reactions for nitro- and azo-compounds, respectively.

Nitro-compounds are found in chemicals such as industrial solvents, insecticides, and food preservatives. Azo-compounds are strongly colored, and are widely used as colorants in foods, cosmetics, pharmaceuticals, and textile and printing industries. The functional properties of nitroreductases and azoreductases are briefly discussed below:

(a) Nitroreductase

Nitroreductase catalyzes the reduction of nitro ($-NO_2$) groups in a wide range of foreign compounds to produce the corresponding hydroxylamines. In the metabolism of nitro compounds, the nitro group is converted to primary amine metabolites, where nitro is initially reduced to nitroso (-NO), then to hydroxylamine (-NHOH), and finally hydroxylamine is converted to primary amine ($-NH_2$).

(b) Azoreductase

Azoreductase catalyzes the reduction reaction of azo-compounds to primary amine metabolites. Azo is initially reduced to hydrazo and then to primary amine. In azoreductase-catalyzed reactions, two equivalents of NAD(P)H are used to reduce one equivalent of the substrate (azo compound). Such a catalytic reaction of azo compounds involves cleavage to yield two amines.

5.2.3 Hydrolytic Enzymes

Hydrolytic enzymes (hydrolases) include carboxylesterases and epoxide hydrolases. Carboxylesterases are found in both the endoplasmic reticulum and the cytosol of the cells. These enzymes are involved in the metabolic activation of various foreign compounds, including drugs, while epoxide hydrolases are present in two forms: soluble and microsomal epoxide hydrolases. They catalyze the hydrolysis of epoxides produced by CYP450-catalyzed reactions.

(a) Carboxylesterase

Carboxylesterase catalyzes the hydrolysis reaction of carboxylic ester to form alcohol and carboxylate. Carboxylesterase isozymes belong to the family of hydrolases. They engage in the hydrolysis of esters and amides, especially acting on carboxylic ester bonds. The hydrolysis of carboxylic amide yields N-hydroxide and carboxylate. Carboxylesterase-catalyzed reactions do not always lead to a detoxification process, since some xenobiotics are converted to chemically active metabolites, such as the conversion of urea derivative to diazonium hydroxide.

(b) Epoxide Hydrolase

Epoxide hydrolase plays an important role in the detoxification of electrophilic epoxides generated from oxidative activation, such as CYP450-catalyzed epoxidation. In some cases, epoxide hydrolases are also considered phase II enzymes.

5.3 Catalytic Reactions

Phase I enzyme-catalyzed reactions may act on different atoms or groups of foreign compounds, such as carbon, nitrogen, oxygen, sulfur, ester, amide, and epoxide. Typical examples of the catalytic actions of activation enzymes on specific groups are presented below. While, detailed descriptions of phase I activation enzyme catalytic reactions are further described in another chapter.

5.3.1 Oxidative Reactions

Oxidative reactions, as described below, emphasize the conversion of specific atoms or groups of foreign compounds to different functional groups in activation reactions catalyzed by phase I enzymes. In the equations presented below, R and Φ represent aliphatic and aromatic derivatives, respectively.

(a) Oxidation at Carbon Atom

Phase I metabolic enzymes that catalyze oxidation reactions include CYP450, alcohol dehydrogenase, aldehyde dehydrogenases, xanthine oxidases, and peroxidases. These oxidation reactions can occur at a carbon atom on an alcohol, aldehyde, or other functional group of a foreign compound.

1. For example, in the reaction catalyzed by CYP450, an oxygen atom is inserted into a C-H bond of an aromatic compound to form phenol, as shown below.

5.3 Catalytic Reactions

$$\Phi\text{-H} + O_2 + \text{NADPH} + H^+ \rightarrow \Phi\text{-OH} + H_2O + \text{NADP}^+$$

While alcohol dehydrogenase-catalyzed oxidation reaction converts aliphatic, aromatic, or cyclic alcohol to aldehyde by removing two hydrogens: one attached to carbon and another attached to oxygen.

$$R\text{-CH}_2\text{OH} + \text{NAD}^+ \rightarrow R\text{-CHO} + \text{NADH} + H^+$$

Further reaction oxidizes aldehyde to form carboxylic acid.

$$R\text{-CHO} + H_2O + \text{NAD}^+ \rightarrow R\text{-COOH} + \text{NADH} + H^+$$

2. In oxidation reactions catalyzed by aldehyde oxidases, an oxygen atom is inserted in C-H bond of aldehyde to yield carboxylic acid, as shown below:

$$R\text{-CHO} + H_2O + O_2 \rightarrow R\text{-COOH} + H_2O_2$$

In peroxidases-catalyzed reactions, alcohol is oxidized to aldehyde by peroxide.

$$R\text{-CH}_2\text{OH} + H_2O_2 \rightarrow R\text{-CHO} + 2\,H_2O$$

(b) Oxidation at Nitrogen Atom

Oxidation reactions can also occur at the nitrogen atom of an amine or amino group of a foreign compound. Enzymatic oxidation reactions at nitrogen atoms include N-hydroxylation and N-oxidation, which are catalyzed by phase I metabolic enzymes such as CYP450 and flavin-containing monooxygenase.

N-hydroxylation involves the substitution of one hydrogen atom in an amino group with a hydroxy group, while N-oxidation involves the addition of oxygen to an amino group. Typical examples of oxidation at a nitrogen atom are shown below:

1. Amine is oxidized to form hydroxyl amine, and imine is oxidized to form oxime in N-hydroxylation, as shown in the following reactions:

$$2\;\Phi\text{-N(H)-CO-CH}_3 + O_2 \rightarrow 2\;\Phi\text{-N(OH)-CO-CH}_3$$

$$2\;R\text{-C(R')=N-H} + O_2 \rightarrow 2\;R\text{-C(R')=N-OH}$$

2. While, in flavin monooxygenases, oxygen atom is added to nitrogen in an amine to form a hydroxyl group.

$$\underset{\substack{|\\ \Phi-N-H}}{CH_3} + O_2 \rightarrow \underset{\substack{|\\ \Phi-N-OH}}{CH_3}$$

(c) Oxidation of Unsaturated Hydrocarbon

Oxidation of an unsaturated hydrocarbon is an important step in foreign compound metabolism. Microsomal CYP450-dependent monooxygenase catalyzes oxidation reaction to convert unsaturated aliphatic or aromatic hydrocarbon to form epoxide. The produced epoxide can undergo further enzymatic reaction catalyzed by epoxide hydrolase to yield hydrodiol.

While unhydrolyzed epoxide can either react with proteins or DNA or form conjugate with glutathione.

1. In CYP450-catalyzed epoxidation, an oxygen atom is inserted into a C=C bond of unsaturated aliphatic or aromatic hydrocarbon to form epoxide, as shown in the following reaction:

$$R-C=C-R' + O_2 \rightarrow \overset{O}{\underset{R-C-C-R'}{/\ \backslash}}$$

2. The produced epoxide can be further converted to dihydrodiol according to the following reaction:

$$\overset{O}{\underset{R-C-C-R'}{/\ \backslash}} + H_2O \rightarrow \underset{\substack{|\\ OH}}{\overset{OH}{\underset{R-C-C-R'}{|}}}$$

5.3.2 Reductive Reactions

A foreign compound containing an azo or nitro group can carry out reduction reaction by interacting with a reducing agent such as NADPH. The reductive reaction catalyzed by azoreductase or nitroreductase occurs at nitrogen atom.

While some enzymes, such as alcohol dehydrogenase, can catalyze either reductive or oxidative reaction, depending on the substrate and conditions. The reductive reaction catalyzed by alcohol dehydrogenase occurs at carbonyl group.

5.3 Catalytic Reactions

Azoreductase, nitroreductase, and alcohol dehydrogenase-catalyzed reductive reactions are presented below.

(a) Reduction at Nitrogen Atom
   1. In azoreductase-catalyzed reaction, N=N in an azo is reduced to form an amine.

$$\Phi-N=N=\Phi' + 2\,NADPH + 2\,H^+ \rightarrow \Phi-NH_2 + H_2N-\Phi' + 2\,NADP^+$$

   2. While, in nitroreductase-catalyzed reaction, a nitro (-NO$_2$) group is reduced to form an amine.

$$R-\Phi-NO_2 + 2\,NADPH + 2\,H^+ \rightarrow R-\Phi-NH_2 + H_2O + 2\,NADP^+$$

(b) Reduction of Carbonyl Group

Alcohol dehydrogenase-catalyzed reaction reduces a carbonyl group to form a hydroxyl group as shown below:

$$R-CO-R' + NADPH + H^+ \rightarrow R-\underset{H}{\overset{OH}{\underset{|}{\overset{|}{C}}}}-R' + NADP^+$$

5.3.3 Hydrolytic Reaction

A substrate containing an ester or amide group can undergo hydrolytic reaction catalyzed by carboxylesterase or hydrolase to produce carboxylic acid as well as alcohol or carboxylic acid and amine. Carboxylesterase is known as serine esterase, because its catalytic site contains a serine residue that participates in the hydrolysis of ester or amide. Enzymatic hydrolysis of amides generally occurs more slowly than that of esters. Hydrolyses of ester and amide are described below:

(a) Hydrolysis of Ester

Carboxylesterase catalyzes hydrolytic reaction that converts an ester to carboxylic acid and alcohol according to the following reaction:

$$\underset{R-C-R'}{\overset{O=C-OC_2H_5}{|}} + H_2O \rightarrow \underset{R-C-R'}{\overset{COOH}{|}} + C_2H_5OH$$

(b) Hydrolysis of Amide

Carboxylesterase also catalyzes hydrolysis reaction to convert an amide to carboxylic acid and amine, as shown in the following reaction;

$$\Phi\text{-CO-N-H} \underset{|}{\overset{R}{}} + H_2O \rightarrow \Phi\text{-COOH} + H_2N\text{-R}$$

Bibliography

Abell CW, Kwan SW (2000) Molecular characterization of monoamine oxidases A and B. Prog Nucleic Acid Res Mol Biol 65:129–156

Beedham C (1997) The role of non-P450 enzymes in drug oxidation. Pharm World Sci 19:255–263

Brash AR (1999) Lipoxygenases: occurrence, functions, catalysis, and acquisition of substrate. J Biol Chem 274:23679–23682

Cashman JR (2002) Human flavin-containing monooxygenase (form 3): polymorphisms and variations in chemical metabolism. Pharmacogenomics 3:325–339

Cashman JR, Zhang J (2006) Human flavin-containing monooxygenases. Annu Rev Pharmacol Toxicol 46:65–100

Chen C-H (2010) Activation and detoxification enzymes: function and implications. Springer, New York

Chen C-H (2020) Xenobiotic metabolic enzymes: bioactivation and antioxidant defense. Springer Nature, Cham

Edmondson DE, Mattevi A, Binda C et al (2004) Structure and mechanism of monoamine oxidase. Curr Med Chem 11:1983–1993

Gershater MC, Cummins I, Edwards R (2007) Role of a carboxylesterase in herbicide bioactivation in Arabidopsis thaliana. J Biol Chem 282:21460–21466

Guengerich FP (1991) Reactions and significance of cytochrome P-450 enzymes. J Biol Chem 266: 10019–10022

Guengerich FP (2002) Cytochrome P450. In: Ioannides C (ed) Enzyme system that metabolize drugs and other xenobiotics. John Wiley, New York

Harini Venkataraman H, den Braver MW, Vermeulen NPE et al (2014) Cytochrome P450-mediated bioactivation of mefenamic acid to quinoneimine intermediates and inactivation by human glutathione S-transferases. Chem Res Toxicol 27(12):2071–2081

Hodek P, Koblihova J, Kizek R et al (2013) The relationship between DNA adduct formation by benzo[a]pyrene and expression of its activation enzyme cytochrome P450 1A1 in rat. Environ Toxicol Pharmacol 36(3):989–996

Ioannides C (2002) Xenobiotic metabolism: an overview. In: Ioannides C (ed) Enzymes systems that metabolise drugs and other xenobiotics. John Wiley & Son, New York

Jelski W, Szmitkowski M (2008) Alcohol dehydrogenase (ADH) and aldehyde dehydrogenase (ALDH) in the cancer diseases. Clin Chim Acta 395(1–2):1–5

Kulkarni AP (2001) Lipoxygenase--a versatile biocatalyst for biotransformation of endobiotics and xenobiotics. Cell Mol Life Sci 58:1805–1825

Meijer J, DePierre JW (1988) Cytosolic epoxide hydrolase. Chem Biol Interact 64:207–249

O'Brien PJ (2000) Peroxidases. Chem Biol Interact 129:113–139

Parkinson A, Ogilvie BW (2008) Biotransformation of xenobiotics. In: Klaassen CD (ed) Casarett & Doull's toxicology: the basic science of poisons. McGraw-Hill

Rittle J, Green MT (2010) Cytochrome P450 compound I: capture, characterization, and C-H bond activation kinetics. Science 330(6006):933–937

Satoh T, Hosokawa M (2006) Structure, function and regulation of carboxylesterases. Chem Biol Interact 162:195–211

Schaich KM (1992) Metals and lipid oxidation. Contemporary issues. Lipids 27(3):209–218

Smith BJ, Curtis JF, Eling TE (1991) Bioactivation of xenobiotics by prostaglandin H synthase. Chem Biol Interact 79(3):245–264

Strolin Benedetti M, Tipton KF (1998) Monoamine oxidases and related amine oxidases as phase I enzymes in the metabolism of xenobiotics. J Neural Transm Suppl 52:149–171

Tipton KF, Benedetti MS (2002) Amine oxidases and the metabolism of xenobiotics. In: Ioannides C (ed) Enzyme system that metabolize drugs and other xenobiotics. John Wiley, New York

Vogel C (2000) Prostaglandin H synthases and their importance in chemical toxicity. Curr Drug Metab 1(4):391–404

Ziegler DM (2002) An overview of the mechanism, substrate specificities, and structure of FMOs. Drug Metab Rev 34:503–511

Phase II Detoxification Enzymes

Phase I enzymes catalyze the activation metabolism of foreign compounds. Their functionalization reactions incorporate a functional group into a foreign compound, resulting in the formation of an intermediate metabolite, which is then detoxified by phase II metabolic reaction. Phase II enzyme-catalyzed reactions serve as a detoxification process by conjugating the incorporated functional group in a foreign compound with a small, endogenous molecule called conjugating ligand.

Phase II enzymes are referred to as detoxification enzymes since they catalyze reactions to deactivate and detoxify foreign compounds. In detoxification process, during phase II metabolism, activation metabolism produces intermediate metabolites that become less reactive and more soluble in water, thus facilitating their excretion from the body through urine or bile. However, for a foreign compound that already possesses a soluble functional group, it can bypass phase I metabolism and directly take part in phase II metabolism.

6.1 Exclusion of Foreign Compounds

Many metabolic intermediates contain highly reactive chemical groups that have the potential to react with cellular components, including proteins, nucleic acids, and lipids. The presence of such chemically active intermediates can lead to adverse health effects and various disease conditions. Phase II conjugation or non-conjugation reactions reduce the reactivity of intermediate metabolites, thus diminishing their potential toxicity to the cells.

In general, conjugation reactions greatly increase the aqueous solubility of foreign compounds, thus facilitating their elimination from the body. Although conjugation reactions generally result in a great increase in the solubility of foreign compounds, there are exceptional cases, such as methylation and acetylation, that produce conjugates with less solubility.

Phase II enzymes mainly belong to the class of transferases, which are responsible for catalytic conjugation reactions in phase II metabolism. Besides conjugation enzymes, a few non-conjugation enzymes also engage in detoxification processes.

6.2 Detoxification Enzymes

Humans are exposed to a wide range of foreign compounds, from chemicals, foods, environments to pharmaceuticals. Enzymatic mechanisms are developed to detoxify and remove these substances. The ability of detoxification enzyme catalytic reaction to detoxify and remove foreign compounds can affect various disease processes. Dairy, lifestyle, and genetic factors can also affect the activities of detoxification enzyme catalytic reactions.

6.3 Conjugation Enzymes

Phase II detoxification enzymes catalyze conjugation and non-conjugation reactions, making metabolites more soluble in water, less reactive towards cell components, and easier to be eliminated in the urine. Conjugation enzymes are composed of a set of transferase enzymes, including uridine-diphosphate-glucuronosyl-transferases, glutathione S-transferases, sulfotransferases, aceyltransferases, methyltransferases, and acyltransferases.

The excretion of conjugates from cells requires ATP-dependent export pumps such as multidrug resistance proteins, which are carried out by phase III metabolism. Phase III metabolism is responsible for the export of conjugates from the cells. Schematic representations of phases I, II, and III metabolisms are shown in Fig. 6.1.

6.3.1 Uridine-Diphosphate-Glucuronosyltransferase

Uridine-diphosphate-glucuronosyltransferase (UGT) is one of the most important transferases that catalyze conjugation reactions in phase II metabolism. The enzyme is located in the endoplasmic reticulum and contains regions of the transmembrane domain. UGT activity is highly dependent on lipids. UGTs are a family of enzymes that catalyze the covalent binding of glucuronic acid to a wide range of lipophilic chemicals.

Glucuronidation reaction involves the transfer of glucuronic acid from uridine-diphosphate-glucuronic acid to a functional group of foreign substrate compound. The liver is the major site of glucuronidation in the living organisms. Other tissues where glucuronidation is present include the kidney, gastrointestinal tract, and lungs. A major consequence of glucuronidation is a significant increase in the aqueous solubility of foreign compounds.

As a major pathway in phase II metabolism, glucuronidation represents one of the most important phase II reactions involving the conversion of lipophilic xenobiotics

6.3 Conjugation Enzymes

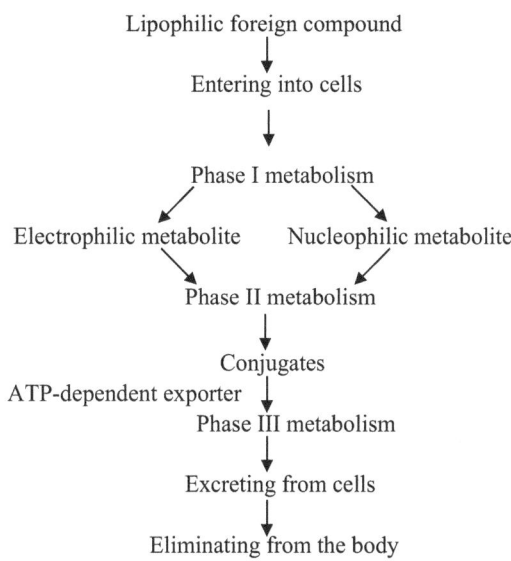

Fig. 6.1 Schematic representations of Phase I, II, and III metabolisms

and their metabolites into hydrophilic conjugates, thus facilitating their excretion. Glucoronidation accounts for about 35% of drug conjugations. The site of glucuronidation is generally an electron-rich nucleophilic atom (O, N, or S).

The substrates for glucuronidation consist of functional groups, such as aliphatic alcohols and phenols, carboxylic acids, or aromatic and aliphatic amines. Glucuronic acid conjugates with the phenolic, carboxylic, or sulfhydryl groups of a substrate to form O-glucuronide and with the amino group of a substrate to form N-glucuronide. Glucoronidation is the primary metabolic reaction for many compounds containing functional groups such as -OH, -COOH, -SH, and -NH2.

Lipophilic foreign compounds catalyzed in phase I reactions are converted into electrophilic or nucleophilic metabolites. Electrophilic metabolites are often conjugated by glutathione S-transferases, while nucleophilic metabolites are mainly conjugated by UGTs, sulfotransferases, and other transferases. UGTs that catalyze conjugation reactions generate products that are more polar, more soluble in water, and more readily available for excretion. After glucuronidation, the produced glucuronides are transported to the kidneys and are available for excretion in the urine.

6.3.2 Glutathione S-Transferase

Glutathione transferase (GST) enzyme families are differentiated into cytosolic enzymes and membrane bound enzymes. Human GST is a multigene family of enzymes that are involved in the metabolism of a wide range of foreign compounds, including arene oxides, unsaturated carbonyls, organic halides, and other xenobiotics.

Glutathione conjugation is an important mechanism in metabolizing electrophilic foreign compounds. Electrophiles are potentially toxic species, since they can bind to nucleophiles such as proteins and nucleic acids, leading to cellular damage and genetic mutation. Foreign compounds that undergo GST-catalyzed conjugation reactions include alkyl and aryl halides, isothiocyanates, unsaturated carbonyls, and epoxides.

GST catalyzes the conjugation of glutathione (GSH) to a variety of foreign compounds. It detoxifies not only electrophiles derived from foreign compounds, but also endogenous electrophiles. Endogenous electrophiles reacting with oxygen are usually the consequence of free radicals, which can also cause damages to cellular components.

Besides serving as a catalyst for conjugation reactions, GST also involves the transfer of glutathione, a reaction that occurs in the cytosol of the liver. Glutathione is synthesized from γ-glutamic acid, cysteine, and glycine. It is abundant in most cells, especially in the liver. Glutathione has the ability to bind many hydrophobic substances, such as bilirubin, steroids, and polycyclic aromatic hydrocarbons.

Glutathione S-transferase catalyzes the conjugation of reduced GSH with a xenobiotic through the formation of a thioether bond between the sulfur atom of GSH and carbon or nitrogen in an electrophilic foreign compound. The conjugate is subsequently metabolized to form cysteine or N-acetylcysteine. Substrates for GST catalyzed reactions commonly exhibit hydrophobic and electrophilic characteristics.

GSTs also play an important role in the detoxification of a broad spectrum of electrophilic metabolites, such as aflatoxin B_1 and benzo[a]pyrene, that may lead to cytotoxicity. GST-catalyzed conjugation reactions inactivate their reactive metabolites and facilitate their excretion from the body.

6.3.3 Sulfotransferase

Sulfotransferase (ST)-catalyzed conjugation reaction is another major pathway of phase II metabolism. Cytosolic sulfotransferases constitute five different families. They catalyze sulfonate conjugation reactions of relatively small lipophilic xenobiotics. A variety of functional groups in xenobiotics are substrates for sulfotransferases, such as hydroxyl group of aliphatic and aromatic compounds.

Sulfonation reaction involves the transfer of a sulfonate group (SO_3^-) from the donor, 3′-phosphoadenosine 5′-phosphosulfate (PAPS), to the acceptor, a nucleophilic group of foreign compounds. PAPS serves as the cofactor of the enzyme for the sulfonation reaction. It is synthesized in tissues to make available an activated form of sulfonate for sulfotransferase-catalyzed reactions. Sulfonation conjugation reaction involves the nucleophilic attack of oxygen or nitrogen atom in a foreign compound on sulfur atom in PAPS, resulting in the cleavage of the phosphosulfate bond. The synthesis of PAPS is dependent on the availability of sulfonate and the activity of the enzymes involved in its synthesis.

Many foreign compounds, such as aromatic and aliphatic alcohols that are involved in O-glucuronidation, also undergo sulfonation. Other compounds involving sulfonation conjugation include aromatic amines such as aniline. Sulfonation generally produces a highly water-soluble sulfuric acid ester. The introduction of negative charge (SO_3^-) to a foreign compound affects not only its aqueous solubility but also its interaction with transport proteins.

Although sulfonation is an effective means of decreasing the toxic activities of foreign compounds, the sulfonate is an electron withdrawing group, and certain sulfonate conjugates are chemically unstable, which can be degraded to form potent electrophilic species. For instance, in drug metabolism, chemically active conjugates may degrade to generate reactive electrophilic cations. Hence, sulfonation affects different physiological processes, including deactivation and bioactivation of foreign compounds.

6.3.4 N-Acetyltransferase

N-acetyltransferases (NAT) are cytosolic enzymes found in the liver and other tissues. NATs catalyze the transfer of the acetyl group of acetyl-coenzyme A to aromatic amine or hydrazine group of a foreign compound, where acetyl-coenzyme A acts as the cofactor of the reaction.

Two functional NAT genes present in humans are NAT1 and NAT2. They are closely related proteins with an active site cysteine residue. There are over 25 allelic variants of NAT1 and NAT2.

NATs are responsible for the metabolism of chemicals that contain aromatic amines or hydrazine. These chemicals are converted to aromatic amide or hydrazide, respectively. In NAT-catalyzed reaction, the acetyl group of acetyl-coenzyme A is initially transferred to a cysteine residue in NAT active site, which is then transferred to the amino group of the substrate such as foreign compound.

Unlike UGT, GST, or ST-catalyzed reactions, acetylation is not associated with increasing aqueous solubility of foreign compounds. The addition of the acetyl group often leads to a conjugate that is less soluble in water because of the covalent binding of the acetyl group to amine or hydrazine group. Nevertheless, acetyl conjugation generally is a detoxification reaction, which results in facilitating the elimination of foreign compounds from the body.

6.3.5 Methyltransferase

Methyltransferases are primarily cytosolic enzymes, but they are also present in the endoplasmic reticulum. Methylation is a minor metabolic route in xenobiotic metabolism. A conjugation reaction catalyzed by methyltransferase involves the transfer of the methyl group attached to the sulfonium ion of S-adenosyl-methionine (SAM) to the functional group of the substrate to form a methyl conjugate. SAM acts as the methyl donor, and the functional group of xenobiotics acts as the acceptor.

Foreign compounds can undergo O-methylation, N-methylation, and S-methylation through SAM dependent conjugation reactions. Pharmacogenetic studies have characterized several methyltransferases, including catechol O-methyltransferase, thiopurine methyltransferase, and histamine N-methyltransferase. The functional groups of a substrate that are involved in methyl conjugation include phenol (O-methylation), aromatic amines (N-methylation), and sulfhydryl-containing groups (S-methylation).

Methyl conjugation usually carries out a detoxification metabolism, resulting in a more rapid elimination of foreign compounds. The methylation of inorganic arsenic to monomethyl-arsonic acid and dimethyl-arsinic acid has been considered to be the major pathway for inorganic arsenic biotransformation and detoxification. Moreover, methyltransferases have been reported to exhibit inheritance variations, which may contribute to individual's variation in arsenic metabolism and arsenic-dependent carcinogenesis.

6.3.6 Acyltransferase

Acyltransferases exist in two distinct forms involving the conjugation of acyl portion of carboxyl group with various amino acids, which occurs in both the liver and kidney. Acyltransferase-catalyzed conjugation reaction carries out the conjugation of carboxylic acid in xenobiotics with acyl-CoA to produce an acyl-CoA thioester, which then reacts with the amino group of an amino acid to form an amide linkage.

Amino-acid conjugation of foreign compounds containing carboxylic acid is considered an alternative to glucuronidation conjugation. Acyltransferase-catalyzed amino acid conjugation of carboxylic acid results in detoxification and elimination of xenobiotics. However, in some cases, glucuronidation of xenobiotics is not simply a detoxification process, since the formation of potentially toxic acylglucuronides could occur.

6.4 Conjugation Enzyme-Catalyzed Reactions

Phase II enzyme-catalyzed reactions can occur at different atoms or groups of foreign compounds, including phenol, epoxide, polyphenol, carboxylic acid, and amino acid. Typical examples of catalytic actions of phase II enzymes include conjugation and non-conjugation reactions. In conjugation enzyme-catalyzed reactions, conjugation can occur at O, N, C, and S atoms in addition to carboxylic acid. These reactions are described below. For general description, R and Φ denote aliphatic and aromatic derivatives, respectively.

6.4.1 Conjugation at O Atom

Oxygen atom of hydroxyl group in phenol is conjugated with glucuronic acid (GA) to form a glucuronide derivative as shown in the following chemical equation:

$$\Phi - OH \ + \ UDP - GA \ \rightarrow \ \Phi - O - GA \ + \ UDP$$

6.4 Conjugation Enzyme-Catalyzed Reactions

6.4.2 Conjugation at N Atom

Conjugation at N atom discussed below includes glucuronic acid (GA) attaching to the N atom of amine, methyl group of S-adenosyl-methionine (SAM) attaching to the N atom of amine, acetyl group of acetyl-CoA binding to the N atom of amine, and sulfonate binding to the N atom of aromatic amine. These chemical reactions are described below:

(a) Glucuronic acid (GA) attaches to the N atom of amine.

Formation of glucuronide conjugation accords to the following equation:

$$\underset{\underset{H}{|}}{\overset{\overset{OH}{|}}{\Phi-N}} + UDP\text{–}GA \rightarrow \underset{\underset{GA}{|}}{\overset{\overset{OH}{|}}{\Phi-N}} + UDP$$

(b) Methyl group of S-adenosyl-methionine (SAM) attaches to the N-atom of amine

Formation of methyl conjugate is presented in the following equation:

$$\Phi\text{–NH–R} + SAM \rightarrow \underset{}{\overset{\overset{CH_3}{|}}{\Phi\text{–N–R}}} + SAH$$

(c) Acetyl group of acetyl-CoA binds to the N atom of amine
In forming acetyl conjugate, acetyl-CoA serves as the cofactor of the enzyme, as shown in the equation below:

$$\Phi - NH_2 + \text{Acetyl-CoA} \rightarrow \Phi - NH - COCH_3 + CoA$$

(d) Sulfonate binds to the N atom of aromatic amine
Formation of sulfonate conjugate accords to the following equation:

$$\Phi\text{–}NH_2 + PAPS \rightarrow \underset{\underset{H}{|}}{\Phi\text{–N–SO}_3H} + PAP$$

6.4.3 Conjugation at C Atom

In glutathione S-transferase (GST)-catalyzed reaction, reduced glutathione (GSH) binds to C atom of epoxide to form a glutathione conjugate according to the following equation:

$$\Phi-\underset{H}{\overset{O}{\overset{/\backslash}{C}-\overset{|}{C}}}-R' \;+\; GSH \;\longrightarrow\; R-\underset{H}{\overset{OH}{\overset{|}{C}}}-\underset{SG}{\overset{|}{C}}-R'$$

6.4.4 Conjugation at S Atom

Conjugation reactions at S atom includes uridine-diphosphate-glucuronosyltransferase and methyltransferase-catalyzed reactions.

(a) Uridine-diphosphate (UDP) - glucuronosyltransferases (GT)-catalyzed reaction

Glucuronic acid binds to the S atom of thio compound to form a glucuronide conjugate, as shown below:

$$R-SH \;+\; UDP-GA \;\longrightarrow\; R-S-GA \;+\; UDP$$

(b) Methyltransferase-Catalyzed Reaction

The methyl group attached to the sulfonium ion in SAM binds to the S atom of thio compound to form methyl conjugate, as presented in the following equation:

$$\Phi-SH \;+\; SAM \;\longrightarrow\; \Phi-S-CH_3 \;+\; SAH$$

6.4.5 Conjugation of Carboxylic Acid

Conjugation of carboxylic acid includes uridine-diphosphate-glucuronosyltransferase and N-acyltransferase-catalyzed reactions.

(a) Uridine-diphosphate-glucuronosyltransferase-Catalyzed Reaction

Glucuronic acid reacts with carboxylic acid to form a glucuronide conjugate, as shown in the following reaction:

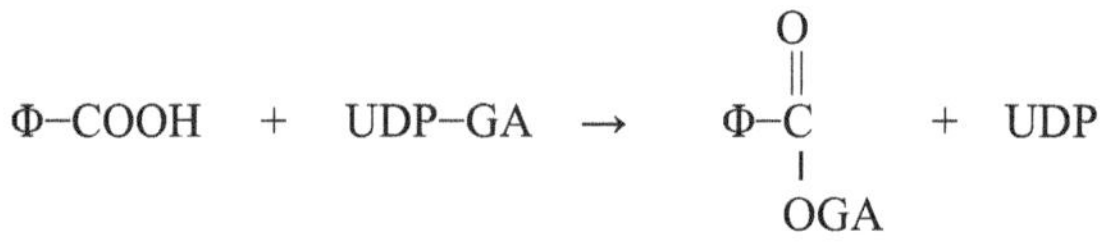

$$\Phi-COOH \;+\; UDP-GA \;\longrightarrow\; \Phi-\underset{OGA}{\overset{O}{\overset{\|}{C}}} \;+\; UDP$$

(b) N-acyltransferase-Catalyzed Reaction

An amino acid reacts with a carboxylic acid group to form an amide as presented below, where Acyl-CoA serves as the cofactor:

$$\Phi\text{-COOH} + \text{HN-CH}_2\text{COOH} \rightarrow \Phi\text{-CO-N-CH}_2\text{COOH} + H_2O$$
$$\quad\quad\quad\quad\quad\quad |\quad\quad\quad\quad\quad\quad\quad\quad\quad\quad\quad\quad |$$
$$\quad\quad\quad\quad\quad\quad R$$

6.4.6 Conjugation at OH Group

Conjugation at OH group involve methyltransferase-catalyzed reaction and oxygen atom of hydroxyl group in alcohol or phenol, as discussed below:

(a) Methyltransferase-Catalyzed Reaction

Oxygen atom of hydroxyl group in phenol is conjugated with methyl group of S-adenosylmethionine to form methyl conjugate, according to the following equation:

$$\Phi\text{-OH} + \text{SAM} \rightarrow \Phi\text{-O-CH}_3 + \text{SAH}$$

where methyl group in SAM binds to O atom in phenol, and SAM and SAH denote S-adenosylmethionine and S-adenosyl-L-homocysteine, respectively.

(b) Oxygen Atom of Hydroxyl Group in Alcohol or Phenol

Alcohol or phenol is conjugated with sulfonate to form a sulfonate derivative, as shown in the following reaction:

$$\Phi\text{-OH} + \text{PAPS} \rightarrow \Phi\text{-O-SO}_3\text{H} + \text{PAP}$$

where PAPS and PAP denote 3-phosphoadenosine 5- phosphosulfate and 3-phosphoadenosine 5- phosphate, respectively.

6.5 Non-conjugation Enzymes

Phase I oxidative reactions can generate quinones and epoxides as metabolite intermediates during the metabolism of foreign compounds. The generated quinone and epoxide intermediates are unstable and chemically reactive. They have been implicated as mutagenic and carcinogenic initiators.

Besides conjugation enzymes, there are also non-conjugation detoxification enzymes, such as quinone reductase and epoxide hydrolase, that play crucial roles in the detoxification of xenobiotics. Non-conjugation enzymes and their associated catalytic reactions are briefly described below.

6.5.1 Quinone Reductase

Quinones are among the toxic products of the oxidative metabolism of aromatic hydrocarbons. Reduction of electrophilic quinones by quinone reductase is an important detoxification pathway. Quinone reductase, a flavoprotein, is one of several phase II enzymes that are involved in the non-oxidative metabolism of a wide variety of foreign compounds.

Quinone reductase displays a broad specificity for structurally diverse hydrophobic quinines, which results in facilitating the metabolism of quinones into readily excreted conjugates.

The enzyme acts on NADH or NADPH with a quinone, where the substrates are NADPH, H^+, and quinone, and the reaction products are $NADP^+$ and semiquinone.

6.5.2 Epoxide Hydrolases

Epoxide hydrolases catalyze hydrolytic reactions for epoxides. Epoxide hydrolases are present in two forms. Soluble epoxide hydrolase is expressed in the cytosol, while microsomal epoxide hydrolase is present in the endoplasmic reticulum. As a typical detoxification enzyme, microsomal epoxide hydrolase exhibits a high expression level in the liver with a broad spectrum of substrates. Epoxide hydrolase is capable of inactivating structurally different, highly reactive epoxides and thus is important in the defense against adverse effects of foreign compounds.

Epoxide hydrolases are essential in the detoxification of foreign compounds that contain unsaturated carbon–carbon bonds. For example, epoxide hydrolase catalyzes the hydration of chemically reactive epoxides to the corresponding dihydrodiol products, which can then be excreted from the body.

6.6 Non-conjugation Enzyme-Catalyzed Reactions

Catalytic reactions involving non-conjugation enzymes, quinone reductase, and epoxide hydrolase, are briefly described below.

6.6.1 Quinone Reductase

NADPH is involved in the reduction of quinone to hydroquinone during quinone reductase catalyzed reaction, according to the following equation:

$$\underset{\substack{\parallel \\ O}}{\overset{\substack{O \\ \parallel}}{\Phi}} + 2\,\underline{NADPH} \rightarrow \underset{\substack{| \\ OH}}{\overset{\substack{OH \\ |}}{\Phi}} + 2\,NADP$$

6.6.2 Epoxide Hydrolase

Epoxide is hydrolyzed to form dihydrodiol in an epoxide hydrolase-catalyzed reaction, as shown below:

$$\underset{\Phi-C-C-R'}{\overset{O}{\diagup \diagdown}} + H_2O \rightarrow \Phi-\underset{OH}{\overset{OH}{C}}-C-R'$$

Bibliography

Arand M, Cronin A, Adamska M, Oesch F (2005) Epoxide hydrolases: structure, function, mechanism, and assay. Methods Enzymol 400:569–588

Benson AM, Hunkeler MJ, Talalay P (1980) Increase of NAD(P)H:quinone reductase by dietary antioxidants: possible role in protection against carcinogenesis and toxicity. Proc Natl Acad Sci USA 177:5216–5220

Chen C-H (2012) Activation and detoxification enzymes: functions and implications. Springer Sciences, New York

Chen C-H (2020) Xenobiotic metabolic enzymes: bioactivation and antioxidant defense. Springer Nature, Cham

Decker M, Arand M, Cronin A (2009) Mammalian epoxide hydrolases in xenobiotic metabolism and signalling. Arch Toxicol 83:297–318

Dekant W, Vamvakas S (1993) Glutathione-dependent bioactivation of xenobiotics. Xenobiotica 23:873–887

Evans D (1992) N-acetyltransferase. In: Kalow W (ed) Pharmacogenetics of drug metabolism. Pergamon, New York

Fretland AJ, Omiecinski CJ (2000) Epoxide hydrolases: biochemistry and molecular biology. Chem Biol Interact 129:41–59

Gamage N, Barnett A, Hempel N et al (2006) Human sulfotransferases and their role in chemical metabolism. Toxicol Sci 90:5–22

Grant DM, Blum M, Meyer UA (1992) Polymorphisms of N-acetyltransferase genes. Xenobiotica 9-10:1073–1081

Hayes JD, Flanagan JU, Jowsey IR (2005) Glutathione transferases. Annu Rev Pharmacol Toxicol 45:51–88

Kato R, Yamazoe Y (1994) Metabolic activation of N-hydroxylated metabolites of carcinogenic and mutagenic arylamines and arylamides by esterification. Drug Metab Rev 26:413–429

King C, Rios G, Green M, Tephly T (2000) UDP-glucuronosyltransferases. Curr Drug Metab 1: 143–161

Klaassen CD, Boles JW (1997) Sulfation and sulfotransferases 5: the importance of 3′--phosphoadenosine 5′-phosphosulfate (PAPS) in the regulation of sulfation. FASEB J 11: 404–418

Liska DJ (1998) The detoxification enzyme systems. Altern Med Rev 3(3):187–198

Mannervik B (1988) Danielson UH.Glutathione transferases--structure and catalytic activity. CRC Crit Rev Biochem 23:283–337

Meech R, Mackenzie PI (1997) Structure and function of uridine diphosphate glucuronosyltransferases. Clin Exp Pharmacol Physiol 24:907–915

Miners JO, Knights KM, Houston JB et al (2006) In vitro-in vivo correlation for drugs and other compounds eliminated by glucuronidation in humans: pitfalls and promises. Biochem Pharmacol 71:1531–1539

Negishi M, Pedersen LG, Petrotchenko E et al (2001) Structure and function of sulfotransferases. Arch Biochem Biophys 390:149–157

Rao PV, Krishna CM, Zigler JS Jr (1992) Identification and characterization of the enzymatic activity of zeta-crystallin from Guinea pig lens. A novel NADPH:quinone oxidoreductase. J Biol Chem 267:96–102

Weinshilboum RM, Otterness DM, Szumlanski CL (1999) Methylation pharmacogenetics: catechol O-methyltransferase, thiopurine methyltransferase, and histamine N-methyltransferase. Annu Rev Pharmacol Toxicol 39:19–52

Wilce MC, Parker MW (1994) Structure and function of glutathione S-transferases. Biochim Biophys Acta 1205(1):1–18

Wildfang E, Zakharyan RA, Aphoshian HV (1998) Enzymatic methylation of arsenic compounds. VI. Characterization of hamster liver arsenite and methylarsonic acid methyltransferase activities in vitro. Toxicol Appl Pharmacol 152:366–375

Wood TC, Salavagionne OE, Mukherjee B et al (2006) Human arsenic methyltransferase (AS3MT) pharmacogenetics: gene resequencing and functional genomics studies. J Biol Chem 281:7364–7373

Talalay P (2000) Chemoprotection against cancer by induction of phase 2 enzymes. Biofactors 12:5–11

Zakharyan RA, Wildfang E, Aphoshian HV (1996) Enzymatic methylation of arsenic compounds. Toxicol Appl Pharmacol 140:77–84

Catalytic Reactions of Activation Enzymes 7

Phase I activation enzymes catalyze oxidation, reduction, and hydrolysis reactions, resulting in the introduction of functional groups to lipophilic foreign compounds and the increase in the water solubility of the parent compounds. The functional groups are introduced by the reactions, such as N-oxidation, S-oxidation, hydroxylation, O- and N-dealkylation, hydrolysis, and epoxidation, where oxidases, reductases, and hydrolases are involved in oxidation, reduction, and hydrolysis reactions.

These above reactions are catalyzed by activation enzymes, including cytochrome P450, prostaglandine H synthase, flavin monooxygenase, amine oxidase, nitroreductase, azoreductase, molybdenum hydroxylase, alcohol dehydrogenase, ribonucleotide reductase, peroxidase, and carboxylesterase. The reactions catalyzed by these enzymes are summarized in the following Table 7.1.

7.1 Cytochrome P450

Cytochrome P450 family (CYP450) is a monooxygenase that incorporates one of two oxygen atoms into the substrate, while another oxygen atom participates in the formation of water. CYP450s play a prominent role in the biotransformation of a great variety of foreign compounds. They contribute to the clearance of drugs more than any other groups in phase I enzymes. Major reactions catalyzed by CYP450s consist of hydroxylation, epoxidation, dehydrogenation, heteroatom oxygenation, and heteroatom dealkylation.

Besides the carbon hydroxylation in the metabolism of sterols and alkanes, CYP450s also involve a variety of other reactions, such as the oxidation of olefins, acetylenes, and polyunsaturated fatty acids. Many drugs are turned into more hydrophilic substances by hydroxylation during metabolic processes that facilitate drug excretion from the body.

Table 7.1 Catalytic reactions of metabolic activation enzymes

Enzyme	Catalytic Reactions
Cytochrone P450	Hydroxylation
	Epoxidation
	Dehydrogenation
	Oxidation
	Dealkylation
Prostaglandin H synthase	Oxidation
	Reduction
Flavin Monooxygenase	Oxidation
Amino oxidase	Oxidation
Nitroreductase	Reduction
Azoreductase	Reduction
Molybdenum	Oxidation
Alcohol dehydrogenase	Reduction
Peroxidase	Oxidation
Carboxyleesterase	Hydrolysis

7.1.1 Hydroxylation of Aliphatic or Aromatic Compounds

Hydroxylation of aliphatic or aromatic compounds involves the oxidation of hydrogen in an alkyl or aromatic group, which results in the formation of alcohol. Hydroxylation is a common process in the metabolism of sterols and alkanes. CYP450-catalyzed hydroxylation reactions are shown in the following equation, where R and Φ represent alkyl and aromatic derivatives, respectively.

$$R\text{-}CH_3 + O_2 + NADPH + H^+ \rightarrow R\text{-}CH_2OH + H_2O + NADP^+$$

$$\Phi-H + O_2 + NADPH + H^+ \rightarrow \Phi-OH + H_2O + NADP^+$$

The above reactions require NADPH and oxygen. Such hydroxylation reactions are catalyzed by CYP450s or oxygenases.

7.1.2 Epoxidation of Ether

An epoxide is a cyclic ether with three-member ring that arises from oxidative metabolism of foreign compounds through enzymatic oxidation processes. Formation of epoxides can occur as CYP450 metabolites of unsaturated carbon-carbon bonds. The resultant epoxides are usually unstable and chemically reactive. A typical example of epoxidation reaction is presented in the following equation:

7.1 Cytochrome P450

$$R\text{-}C\text{=}C\text{-}R' + O_2 + NADPH + H^+ \rightarrow R\text{-}\overset{\overset{O}{/\,\backslash}}{C}\text{-}C\text{-}R' + H_2O + NADP^+$$

where R and R' represent aliphatic derivatives.

7.1.3 Dehydrogenation of Alcohol or Aldehyde

CYP450-catalyzed dehydrogenation reaction results in the oxidation of alcohol to form aldehyde or aldehyde to form carboxylic acid. Typical examples of dehydrogenation of alcohol and aldehyde are presented in the following equations:

$$R-CH_2OH + O_2 + NADPH + H^+ \rightarrow RCH{=}O + 2H_2O + NADP^+$$

$$R-CHO + O_2 + NADPH + H^+ \rightarrow RCOOH + H_2O + NADP^+$$

where R denotes an aliphatic derivative.

7.1.4 Oxidation of N - or S - Compound

Oxygenation reactions involving N- or S-compounds are commonly seen with amines and sulfides. Amines are oxidized to hydroxyl amines, and thioethers are oxidized to sulfoxides. Typical equations for N- or S-oxidation reactions are shown below:

$$\Phi-NH_2 + O_2 + NADPH + H^+ \rightarrow \Phi-\underset{H}{N\text{-}OH} + H_2O + NADP^+$$

$$\Phi-S-CH_3 + O_2 + NADPH + H^+ \rightarrow \Phi-\underset{CH_3}{S{=}O} + H_2O + NADP^+$$

where Φ represents an aromatic derivative.

7.1.5 Dealkylation of Ether, Amide, or Carboxylic Acid

The cleavage of ether, amide or carboxylic acid is a common CYP450-catalyzed reaction. Dealkylation of ether, amide, or carboxylic acid is presented in the following equations:

$$R-O-R' + O_2 + NADPH + H^+ \rightarrow ROOH + HOR' + NADP^+$$

$$\underset{\underset{NH_2}{|}}{R-C=O} + O_2 + NADPH + H^+ \rightarrow RCOOH + NH_3OH + NADP^+$$

$$\underset{\underset{OH}{|}}{R-C=O} + O_2 + NADPH + H^+ \rightarrow HOOH + HCOOH + NADP^+$$

where R and R' denote aliphatic derivatives.

7.1.6 Oxidation of Carbon on Aromatic Ring

A typical example of CYP450-catalyzed oxidation of a carbon on aromatic ring is shown in the following chemical reaction, where X represents H, N, or S attaching to a carbon on the aromatic ring, and Φ denotes an aromatic derivative.

$$\Phi-X + O_2 + NADPH + H^+ \rightarrow \Phi-O-X + NADP^+ + H_2O$$

7.1.7 Activation of Benzo[a]pyrene

Benzo[a]pyrene (BaP) is a human carcinogen that requires metabolic activation prior to reaction with DNA. CYP1A1 is the most important hepatic or intestinal enzyme in activation of BaP. An induction of CYP1A enzyme activity can result in an increase in formation of BaP-DNA adducts.

7.2 Prostaglandin H Synthase

Prostaglandin H synthase (PGHS) is an enzyme for the oxidation of foreign compounds. Its enzymatic activities can oxidize foreign compounds to form reactive intermediates. Many xenobiotics and carcinogens are oxidized in vitro by PGHS in the presence of arachidonic acid or lipid peroxides. The oxidation of foreign compounds by PGHS often forms metabolites or adducts to cellular macromolecules.

Especially, PGHS plays a role in the bioactivation of polycyclic aromatic hydrocarbons and aromatic amines. It is also the enzyme in arachidonate metabolism that forms prostanoids, such as prostaglandins, prostacyclins, and thromboxanes. PGHS catalyzes the oxidation of arachidonic acid in the reactions that involve cyclooxygenase and peroxidase activities.

PGHS also catalyzes the reduction of hydroperoxyl endoperoxide to hydroxy endoperoxide, where a variety of chemicals can serve as electron donors, such as phenolic compounds, aromatic amines, and polycyclic aromatic hydrocarbons. Since PGHS can oxidize many foreign compounds, it has been suggested to serve as an alternative metabolic activation enzyme to CYP450 isoenzymes, particularly in tissues low in monooxygenase activity.

7.3 Flavin Monooxygenase

Monooxygenases are involved in the oxidation of numerous organic compounds containing nitrogen, sulfur, or phosphorus to form oxides of nitrogen, sulfur, or phosphorus. As oxidative enzymes, flavin-containing monooxygenases exhibit functions that overlap with CYP450s. However, unlike CYP450s, flavin monooxygenase-catalyzed reactions utilize flavin adenosine dinucleotide as the coenzyme.

A typical flavin-containing monooxygenase-catalyzed fatty acid oxidation reaction is shown in the following equation:

$$R-CH_2-CH_2-CO-S-CoA \;+\; FAD \;\rightarrow\; R-CH=CH-CO-S-CoA \;+\; FADH_2$$

where CH_2–CH_2 bond is oxidized to form $CH=CH$.

7.4 Amine Oxidase

Amine oxidases catalyze the oxidation of amines in the metabolism of foreign compounds. The basic reaction is the oxidative cleavage of the α-H in aliphatic or aromatic amines. A typical equation of amine oxidase-catalyzed reaction is shown as the following:

$$RCH2\text{-}NR'R'' \;+\; H_2O \;+\; O_2 \;\rightarrow\; RCHO \;+\; H\text{-}NR'R'' \;+\; H_2O_2$$

The products of the above reaction are aldehyde, ammonia, and hydrogen peroxide. The resultant hydrogen peroxide is the source of hydroxyl radical (·OH). While, the produced aldehyde may be further metabolized by aldehyde oxidase or aldehyde reductase to form carboxylic acid or alcohol according to the following equations:

$$RCHO \;+\; NAD \;+\; H_2O \;\rightarrow\; RCOOH \;+\; NADH_2$$

$$RCHO + NADH_2 \rightarrow RCH_2OH + NAD$$

7.5 Nitroreductase

Nitroreductase catalyzes the reduction of nitro compounds to form primary amine metabolites, according to the following sequential reactions:

$$RNO_2 + NADPH + H^+ \rightarrow RNO + NADP^+ + H_2O$$

$$RNO + NADPH + H^+ \rightarrow RNHOH + NADP^+$$

$$RNHOH + NADPH + H^+ \rightarrow RNH_2 + NADP^+ + H_2O$$

where the nitro group (-NO$_2$) is initially reduced to nitroso (-NO), then to hydroxylamine (-NHOH), and finally to primary amine (-NH$_2$).

7.6 Azoreductase

Azoreductase catalyzes the reduction of azo compounds, where Azo is initially reduced to hydrazo and finally to primary amine according to the following reactions:

$$\Phi-N=N-\Phi' + NADPH + H^+ \rightarrow \Phi-NH-NH-\Phi' + NADP^+$$

$$\Phi-NH-NH-\Phi' + NADPH + H^+ \rightarrow \Phi-NH_2 + \Phi'-NH_2 + NADP^+$$

where Φ and Φ' denote aromatic derivatives, which may be replaced with aliphatic derivative (R).

7.7 Molybdenum Hydroxylase

Molybdenum hydroxylase exhibits oxidase activities toward a variety of heterocyclic compounds and aldehydes. Among the molybdenum hydroxylase families are aldehyde oxidase and xanthine oxidase. They are important in the metabolism of drugs and other foreign compounds. Xanthine oxidase plays an important role in the catabolism of purines, where the oxygen atom inserted into the substrate is from water rather than molecular oxygen.

Xanthine oxidase also catalyzes the oxidation of hypoxanthine to xanthine. The enzyme further catalyzes the oxidation of xanthine to uric acid. The overall reaction is the oxidation of hypoxanthine to form uric acid, as shown in the equation below.

[Chemical reaction: hypoxanthine + H₂O + O₂ → xanthine + H₂O₂]

7.8 Alcohol Dehydrogenase

Alcohol dehydrogenase catalyzes the conversion of primary or secondary alcohol to aldehyde or ketone. Alcohol dehydrogenase-catalyzed reaction requires NAD^+ as a coenzyme. A typical example of alcohol dehydrogenase-catalyzed reaction is presented in the equation below.

$$RCH_2OH + NAD^+ \rightarrow RCHO + NADH + H^+$$

The produced aldehyde is usually toxic, which is further oxidized to acid before its excretion.

7.9 Ribonucleotide Reductase

Ribonucleotide reductase (RNR) is a multi-subunit enzyme responsible for catalyzing the rate-limiting step in the production of deoxyribonucleotides, which is essential for DNA synthesis and repair. Ribonucleotide reductases transform RNA to DNA building blocks by catalyzing the substitution of the OH-group of a ribonucleotide with hydrogen by a mechanism involving protein radicals.

Conformational transitions induced by the nucleotide binding determine the regulation of substrate specificity. A complicated interplay among gene activation, enzyme inhibition, and protein degradation regulates the appropriate amount of deoxynucleotides for DNA replication and repair.

7.10 Peroxidase

Peroxidases are a large family of enzymes, such as horseradish peroxidase and cytochrome c peroxidase. Peroxidases catalyze the conversion of peroxides to form alcohols. A typical peroxidase-catalyzed reaction is shown in the equation below:

$$RO-OR' + H^+ + NADH \rightarrow ROH + R'OH + NAD^+$$

where RO-OR' represents hydrogen peroxide or organic hydroperoxides, such as lipid peroxides.

7.11 Carboxylesterase

Carboxylesterase catalyzes the hydrolysis reaction involving the carboxylic ester bond. The reaction proceeds as shown in the following equation:

$$R-COOR' + H_2O \rightarrow R'OH + RCOO^- + H^+$$

where R − COOR' represents a carboxylic ester, and R and R' denote aliphatic derivatives. The reaction products are alcohol and carboxylate.

Bibliography

Beedham C (1985) Molybdenum hydroxylases as drug-metabolizing enzymes. Drug Metab Rev 16:119–156

Beedham C (1998) Molybdenum hydroxylases. In: Gorrod JW et al (eds) Metabolism of xenobiotics. Taylor and Francis, London and New York

Benedetti MS (2001) Biotransformation of xenobiotics by amine oxidases. Fundam Clin Pharmacol 15:75–84

Benedetti MS, Dostert P (1994) Contribution of amine oxidases to the metabolism of xenobiotics. Drug Metab Rev 26:507–535

Cashman JR (1998) Stereoselectivity in S- and N-oxygenation by the mammalian flavin-containing and cytochrome P-450 monooxygenases. Drug Metab Rev 30:675–707

Cashman JR (2000) Human flavin-containing monooxygenase: substrate specificity and role in drug metabolism. Curr Drug Metab 1:181–191

Cashman JR, Zhang J (2006) Human flavin-containing monooxygenases. Annu Rev Pharmacol Toxicol 46:65–100

Chen C-H (2020) Xenobiotic metabolic enzymes: bioactivation and antioxidant defense. Springer Nature, Cham

Christofferson A, Wilkie J (2009) Mechanism of CB1954 reduction by Escherichia coli nitroreductase. Biochem Soc Trans 37:413–418

Dawson JH (1988) Probing structure-function relations in heme-containing oxygenases and peroxidases. Science 240:433–439

de Montellano O, Paul R, de Montellano O (2005) Cytochrome P450: structure, mechanism, and biochemistry. Kluwer Academic/Plenum Publishers, New York

Degen GH (1993) Prostaglandin-H synthase containing cell lines as tools for studying metabolism and toxicity of xenobiotics. Toxicology 82(1–3):243–256

Edmondson DE, Mattevi A, Binda C et al (2004) Structure and mechanism of monoamine oxidase. Curr Med Chem 11:1983–1993

Guengerich FP (1991) Reactions and significance of cytochrome P-450 enzymes. J Biol Chem 266:10019–10022

Guengerich FP (1996) The chemistry of cytochrome P450 reactions. In: Ioannides C (ed) Cytochrome P450: metabolic and toxicological aspects. CRC Press, Boca Raton

Guengerich FP, Shimada T (1991) Oxidation of toxic and carcinogenic chemicals by human cytochrome P-450 enzymes. Chem Res Toxicol 4:391–407

Guengerich FP, Peterson LA, Böcker RH (1988) Cytochrome P-450-catalyzed hydroxylation and carboxylic acid ester cleavage of Hantzsch pyridine esters. J Biol Chem 263:8176–8183

Heikkila RE, Kindt MV, Sonsalla PK et al (1988) Importance of monoamine oxidase a in the bioactivation of neurotoxic analogs of 1-methyl-4-phenyl-1,2,3,6-tetrahydropyridine. Proc Natl Acad Sci USA 85:6172–6176

Hodek P, Koblihova J, Kizek R et al (2013) The relationship between DNA adduct formation by benzo[a]pyrene and expression of its activation enzyme cytochrome P450 1A1 in rat. Environ Toxicol Pharmacol 36(3):989–996

Hodgson E, Levi PE (1992) The role of the flavin-containing monooxygenase (EC 1.14.13.8) in the metabolism and mode of action of agricultural chemicals. Xenobiotica 22:1175–1183

Ioannides C (ed) (2002) Enzyme systems that metabolise drugs and other xenobiotics. John Wiley, New York

Ito K, Nakanishi M, Lee WC et al (2008) Expansion of substrate specificity and catalytic mechanism of azoreductase by X-ray crystallography and site-directed mutagenesis. J Biol Chem 283: 13889–13896

Koder RL, Haynes CA, Rodgers ME et al (2002) Flavin thermodynamics explain the oxygen insensitivity of enteric nitroreductases. Biochemistry 41:14197–14205

Krenitsky TA, Neil SM, Elion GB, Hitchings GH (1972) A comparison of the specificities of xanthine oxidase and aldehyde oxidase. Arch Biochem Biophys 150:585–599

Kulkarni AP, Hodgson E (1984) The metabolism of insecticides: the role of monooxygenase enzymes. Annu Rev Pharmacol Toxicol 24:19–42

Meunier B, de Visser SP, Shaik S (2004) Mechanism of oxidation reactions catalyzed by cytochrome p450 enzymes. Chem Rev 104:3947–3980

Nordlund P, Reichard P (2006) Ribonucleotide reductases. Annu Rev Biochem 75:681–706

O'Brien PJ (2000) Peroxidases. Chem Biol Interact 129:113–139

Parkinson A, Ogilvie BW (2008) Biotransformation of xenobiotics. In: Klaassen CD (ed) Casarett and Doull's toxicology: the basic science of poisons. McGrawHill, New York

Rafii F, Cerniglia CE (1995) Reduction of azo dyes and nitroaromatic compounds by bacterial enzymes from the human intestinal tract. Environ Health Perspect 103(suppl 5):17–19

Rajagopalan KV (1980) Xanthine oxidase and aldehyde oxidase. In: Jakoby WB (ed) Enzymatic basis of detoxication. Academic Press, New York

Satoh T, Hosokawa M (1998) The mammalian carboxylesterases: from molecules to functions. Annu Rev Pharmacol Toxicol 38:257–288

Satoh T, Hosokawa M (2006) Structure, function and regulation of carboxylesterases. Chem Biol Interact 162:195–211

Seitz HK, Oneta CM (1998) Gastrointestinal alcohol dehydrogenase. Nutr Rev 56:52–60

Senter PD, Marquardt H, Thomas BA et al (1996) The role of rat serum carboxylesterase in the activation of paclitaxel and camptothecin prodrugs. Cancer Res 56:1471–1474

Shimada T, Martin MV, Pruess-Schwartz D et al (1989) Roles of individual human cytochrome P-450 enzymes in the bioactivation of benzo(a)pyrene, 7,8-dihydroxy-7,8 dihydrobenzo(a)pyrene, and other dihydrodiol derivatives of polycyclic aromatic hydrocarbons. Cancer Res 49: 6304–6312

Smith BJ, Curtis JF, Eling TE (1991) Bioactivation of xenobiotics by prostaglandin H synthase. Chem Biol Interact 79(3):245–264

Strolin Benedetti M, Tipton KF (1998) Monoamine oxidases and related amine oxidases as phase I enzymes in the metabolism of xenobiotics. J Neural Transm Suppl 52:149–171

Uetrecht J (2003) Bioactivation. In: Lee JS, Obach S, Fisher MB (eds) Drug metabolizing enzymes. Marcel Dekker, New York

Vasiliou V, Pappa A, Estey T (2004) Role of human aldehyde dehydrogenases in endobiotic and xenobiotic metabolism. Drug Metab Rev 36:279–299

Vogel C (2000) Prostaglandin H synthases and their importance in chemical toxicity. Curr Drug Metab 1(4):391–404

Wong LL (1998) Cytochrome P450 monooxygenases. Curr Opin Chem Biol 2:263–268

Ziegler DM (1988) Flavin-containing monooxygenases: catalytic mechanism and substrate specificities. Drug Metab Rev 19:1–32

Catalytic Reactions of Detoxification Enzymes

8

In metabolic biotransformation process, foreign compounds are converted into more polar products to facilitate their elimination from the body. As discussed above, phase I activation metabolism involves functionalization reactions, where metabolic reactions catalyzed by activation enzymes introduce functional groups to foreign compounds, resulting in greatly increasing the solubility of parent compounds. The modified lipophilic foreign compounds then undergo phase II enzyme-catalyzed reactions to detoxify metabolites and facilitate their excretion from the body.

Phase II detoxification reactions are a major defense mechanism against reactive intermediates of metabolites generated during phase I activation reactions. The enzymes responsible for metabolic reactions include conjugation and non-conjugation enzymes. Conjugation enzymes are such as uridine 5′-diphosphoglucuronosyl transferase, glutathione S-transferase, sulfotransferase, arylamine acetyltransferase, and methyltransferase. Sulfation, glucuronidation, and glutathione conjugation are the three most prevalent classes of phase II metabolism. In addition, quinone reductase and epoxide hydrolase are involved in non-conjugation reactions.

In conjugation reactions, the donor transfers an ionic or non-ionic group to the functional group of the acceptor. Donor compounds include uridine-diphospho-glucuronic acid, glutathione, phosphoadenosine phosphosulfate, adenosylmethionine, acetyl-coenzyme A, glycine, and glutamine. Functional groups of the acceptor include phenol, epoxide, polyphenol, carboxylic acid, and amino acid.

Kinetic analysis of conjugation by hepatic microsomes indicated that the differences in excreted metabolites reflected the differences in enzymatic activities. Typical functional groups of substrates in conjugation reactions are listed in Table 8.1.

Table 8.1 Functional groups of substrates in conjugation reactions

Phase II enzyme	Conjugation reaction	Functional group of substrate
UDP-glucuronyltransferase	Glucuronidation	-OH, SH, -NH2, CH2, -COOH, NHOH
Glutathione S-transferase	Glutathione	Epoxide, halide, -NO2
Sulfotransferases	Sulfonation	-OH, -NH2
N-aceytltransferases	Acetylation	-OH, -NH2, -SO2NH2
Methyltransferases	Methylation	-OH, -NH2, -SH

$$\text{Lipophilic xenobiotic} \xrightarrow[\text{CYP450}]{\text{Phase I}} \text{Nucleophilic metabolite} \xrightarrow[\text{UGT}]{\text{Phase II}} \text{Glucuronide conjugate}$$

$$\xrightarrow[\text{MRP}]{\text{Phase III}} \text{Excretion}$$

Fig. 8.1 Nucleophilic metabolite glucuronide conjugation

8.1 Conjugation Reactions

Conjugation reactions include those reactions catalyzed by uridine-diphospho-glucuronosyl transferase, glutathione S-transferase, sulfotransferase, arylamine acetyltransferase, and methyltransferase. For instance, metabolic conjugation activity was examined by analyzing pyrene, a polycyclic aromatic hydrocarbon metabolite. The major metabolites are pyrene glucuronide and pyrene sulfate.

Conjugation reactions catalyzed by uridine diphospho-glucuronosyl transferase, glutathione S-transferase, sulfotransferase, acyltransferase, N-acetyltransferase, and methyltransferase are discussed below:

8.1.1 Uridine Diphospho-Glucuronosyl Transferase

In phase II metabolism, nucleophilic metabolites converted by phase I enzymes often undergo conjugation reactions catalyzed by uridine 5′-diphospho-glucuronosyl transferase (UGT) to form glucuronide conjugates. Glucuronidation forms a variety of O-, N-, S-, and C- containing glucuronides.

The produced conjugates are more polar and easier excreted than the parent compound. The excretion of glucuronide conjugates from the cells is carried out by ATP-dependent export pumps, such as multidrug resistant proteins. Glucuronidation of foreign compounds occurs in the liver, intestine, and kidney. The produced glucuronide conjugates are usually excreted in urine and bile.

Glucuronidation is an important step in the metabolism of aromatic amines. A schematic description of activation, nucleophilic metabolite, glucuronidation, and excretion is described in Fig. 8.1.

Furthermore, UGT catalyzes the transfer of glucuronic acid (GA) from uridine 5′-diphospho-glucuronic acid (UDP-GA) to a foreign compound that contains oxygen, nitrogen, sulfur, or carboxyl functional group. O-, N-, S-, and C- glucuronide conjugation reactions are shown in the following equations:

$$R-OH + UDP-GA \rightarrow UDP + R-O-GA$$

$$R-SH + UDP-GA \rightarrow UDP + R-S-GA$$

$$R-NH_2 + UDP-GA \rightarrow UDP + R-NH-GA$$

$$R-CHO + UDP-GA \rightarrow UDP + R-CO-GA$$

$$R-COOH + UDP-GA \rightarrow UDP + R-CO(OGA)$$

$$R-NHOH + UDP-GA \rightarrow UDP + R-N(OH)GA$$

Where R denotes an aliphatic derivative and Φ a aromatic derivative. The functional group of a substrate is composed of hydroxyl (-OH), thiol (-SH), amine (-NH_2), carbonyl (-C=O), carboxylic (-COOH), or hydroxylamine (-NHOH).

8.1.2 Glutathione S-Transferase

Glutathione S-transferase (GST) contributes to the metabolism of foreign compounds by catalyzing reactions that conjugate glutathione (GSH) with electrophilic metabolites derived from phase I metabolism. Conjugation with GSH occurs via a sulfhydryl group to the electrophilic center on the foreign compound.

GST plays an important role in the detoxification of a broad range of toxic foreign compounds, particularly those that may lead to cytotoxicity or mutagenic events (aflatoxin B_1 and benzo(a)pyrene). The resulting conjugates generally are less reactive and more water-soluble and are ready for excretion in urine or bile. A schematic outline of activation, conjugation, and excretion is described in Fig. 8.2.

The family of GST isozymes catalyzes the conjugation of GSH with an electrophilic substrate to form a thioester bond between the sulfur atom of GSH and the substrate. Foreign compounds that undergo GST-catalyzed conjugation reactions include alkyl- and aryl-halides, epoxides, isothiocyanates, unsaturated carbonyls, and nitro compounds. Moreover, metabolic reactions involving glutathione

$$\text{Xenobiotic} \xrightarrow[\text{CYP450}]{\text{Phase I}} \text{Metabolite} \xrightarrow[\text{GST}]{\text{Phase II}} \text{GSH conjugate}$$

$$\xrightarrow[\text{MRP}]{\text{Phase III}} \text{Excretion}$$

Fig. 8.2 Excretion of electrophilic metabolite through GSH conjugation

conjugation with epoxide, aliphatic halide, or nitro compounds are also presented in the following equations, where R denotes an aliphatic derivative and Φ an aromatic derivative.

$$R-\underset{\underset{O}{\diagdown\diagup}}{C-C} \quad + \quad GSH \quad \rightarrow \quad R-\underset{\underset{GS}{|}\;\underset{OH}{|}}{C-C}$$

$$R-Cl \quad + \quad GSH \quad \rightarrow \quad GS-R \;+\; H^+ \;+\; Cl^-$$

$$R-NO_2 \quad + \quad GSH \quad \rightarrow \quad GS-R \;+\; H^+ \;+\; NO_2^-$$

8.1.3 Sulfotransferase

Sulfonation is an important conjugation reaction in phase II metabolism. Sulfoconjugation renders a compound more hydrophilic to aid its excretion. Sulfotransferase-catalyzed reaction conjugates the functional group of electrophilic metabolites with sulfo moiety, where the sulfonate group is transferred from a donor molecule (cofactor of enzyme) to an acceptor molecule such as alcohol or amine. Sulfotransferase-catalyzed reactions are presented below for alcohols or phenols, and aliphatic or aromatic amines.

$$R-OH \quad + \quad PAPS \quad \rightarrow \quad R-O-SO_3H \quad + \quad PAP$$

$$R-NH_2 \quad + \quad PAPS \quad \rightarrow \quad R-NH-SO_3H \quad + \quad PAP$$

where R denotes an aliphatic derivative or Φ an aromatic derivative. The most common sulfate donor is 3′-phosphoadenosine-5′-phosphosulfate (PAPS). PAPS denotes 3-phosphoadenosine 5-phosphate. Sulfonate conjugation is the most important pathway in the metabolism of phenols. The major enzyme responsible for xenobiotic sulfonation is the widely expressed cytosolic sulfotransferase SULT1A1.

Sulfonation reaction involves the transfer of sulfonate group ($-SO_3^-$) from PAPS to amine, hydroxylamine, or alcohol. Sulfonation conjugate formation is known to occur with aromatic or aliphatic amines, phenols, as well as primary, secondary, and tertiary alcohols. Sulfonate conjugates are excreted predominately in the urine. Similar to Fig. 8.2 for glutathione S-transferase, the addition of a sulfonate group also facilitates the excretion of a foreign compound from the cells, which is also carried out by multidrug resistance proteins.

8.1 Conjugation Reactions

8.1.4 Acyltransferase

Amino acid conjugation is an important pathway in the metabolism of carboxyl acid-containing foreign compounds, for examples, benzoic acid with glycine, phenylacetic acid with glutamine, xanthurenic acid with serine, and 4-nitrobenzoic acid with arginine. Acyltransferase-catalyzed conjugation reactions act upon the acyl group (R-C=O) of the carboxylic acid (R-COOH) of foreign compounds, leading to the formation of amides, esters, or a peptide bond between the acyl group of xenobiotic and the amino group of endogenous compounds.

Acyl conjugation reactions require an initial activation of xenobiotic to a CoA derivative, which is catalyzed by acyl CoA ligase. The resulting acyl CoA subsequently reacts with an amino acid, giving rise to acylated amino acid conjugate and CoA. Thus, acyl conjugation reaction occurs in two steps: the initial activation of the carboxyl group to yield reactive acyl-CoA thioester, followed by the transfer of acyl to the amino group of an amino acid.

Acyltransferase-catalyzed conjugation reactions are shown in the following chemical equations, where Φ denotes the aromatic portion of a foreign compound such as benzoic, phenylacetic, or xanthurenic acid, and R represents the side chain portion of an amino acid such as glycine, serine, or arginine.

Step 1: $\Phi - COOH\ +\ Acyl\text{-}CoA\ \rightarrow\ \Phi - CO(CoA\text{-}Acyl)\ +\ H_2O$

Step 2: $\Phi\text{-}CO(CoA\text{-}Acyl)\ +\ H_2N\text{-}\underset{R}{\overset{H}{C}}\text{-}COOH\ \rightarrow\ \Phi\text{-}CO\text{-}(HN\text{-}\underset{R}{\overset{H}{C}}\text{-}COOH)\ +\ Acyl\text{-}CoA$

Overall reaction: $\Phi\text{-}COOH\ +\ H_2N\text{-}\underset{R}{\overset{H}{C}}\text{-}COOH\ \rightarrow\ \Phi\text{-}CO\text{-}(HN\text{-}\underset{R}{\overset{H}{C}}\text{-}COOH)\ +\ H_2O$

Amino acid conjugation is an alternative conjugation process for carboxylic acid-containing xenobiotics. It can occur in the liver and kidney. The resulting amino acid conjugates are generally excreted from the body by urinary elimination.

8.1.5 N-Acetyltransferase

Acetylation couples an amino group with the acetyl moiety, which results in the formation of acetylated derivatives that are generally less water-soluble than the parent compound. In N-acetyltransferase catalytic reaction, the transfer of an acetyl group from acetyl-CoA (AcCoA) to the terminal nitrogen of arylamine contains an aromatic hydrocarbon with at least one amine group attached to it. N-acetyltransferase has been shown to be important in the detoxification of drugs.

Acetyltransferase-catalyzed reaction occurs in two steps: the enzyme is acetylated by AcCoA and then the acetyl group (the donor) is transferred to the acceptor, such as arylamine.

Acetyl conjugation reactions are shown in the following two-step equations:

$$\text{First step:} \quad \text{AcCoA} + \text{NAT} \rightarrow \text{NAT} - \text{AcCoA}$$

$$\text{Second step:} \quad \text{NAT-AcCoA} + \Phi\text{-NH}_2 \rightarrow \Phi\text{-N(H)-COCH}_3 + \text{NAT} + \text{CoA}$$

$$\text{Overall reaction} \quad \Phi\text{-NH}_2 + \text{AcCoA} \rightarrow \Phi\text{-N(H)-COCH}_3 + \text{CoA}$$

where NAT represents N-acetyltransferase. $\Phi\text{-NH}_2$ denotes arylamine. Ac, CoA, and AcCoA represent acetyl group ($-COCH_3$), coenzyme A, and acetyl CoA, respectively.

In the case of arylhydroxylamine as the acceptor, the overall reaction is as follow

$$\Phi\text{-NHOH} + \text{AcCoA} \rightarrow \Phi\text{-N(OH)-COCH}_3 + \text{CoA}$$

8.1.6 Methyltransferase

Methyltransferase catalyzes the reaction that transfers the methyl group from the donor to the acceptor (substrate). The methyl donor is the reactive methyl group, which is bound to the sulfur in S-adenosyl methionine (SAM). SAM is the cofactor of methyltransferase. SAM - dependent methyltransferases act on a wide variety of target molecules, such as protein and DNA.

Methylation occurs on nucleic bases in DNA and amino acids in proteins. Hydroxyl (-OH), amino ($-NH_2$) and thiol (-SH) groups can be metabolized through methylation. Methyltransferase - catalyzed reactions involving the transfer of methyl group to N, O, or S- nucleophiles are shown in the following chemical equations:

$$\text{RCH}_2\text{NH}_2 + \text{SAM} \rightarrow \text{RCH}_2\text{N(H)-CH}_3 + \text{SAH}$$

$$\Phi - OH + SAM \rightarrow \Phi - O - CH_3 + SAH$$

$$HS\text{-}CH_2OH + SAM \rightarrow CH_3 - S\text{-}CH_2OH + SAH$$

where R and Φ denote aliphatic and aromatic portions of a foreign compound. SAM and SAH denote S-adenosylmethionine and S-adenosyl-L-homocysteine, respectively.

8.2 Non-conjugation Reactions

Non-conjugation reactions include quinone reductase and epoxide hydrolase-catalyzed reactions. Quinones are among toxic products of CYP450 oxidative metabolism of aromatic hydrocarbons. Epoxides are ethers that contain a three-member ring. They exhibit reactivity due to the highly polarized oxygen-carbon bonds in addition to a highly strained ring.

8.2.1 Quinone Reductase

Quinone reductase is a phase II non-conjugation enzyme that exhibits a broad specificity for structurally diversified quinones. The reduction of electrophilic quinones catalyzed by quinone reductase is an important detoxification pathway. Reactions catalyzed by quinone reductase usually utilize NADH or NADPH as a source of reductant.

Quinone reductase-catalyzed reduction reaction of quinone to phenol contains two -OH groups, which is shown in the following chemical equation:

$$O = \Phi = O + NADH + H^+ \rightarrow HO - \Phi - OH + NAD^+$$

where Φ denotes an aromatic derivative. $O = \Phi = O$ such as in P-benzoquinone is reduced to $O = \Phi - OH$ as in semiquinone, and then to $HO - \Phi - OH$ as in hydroquinone.

8.2.2 Epoxide Hydrolase

Some reactive epoxides are responsible for electrophilic reactions with critical biological targets such as proteins and DNA, leading to carcinogenic effects. Epoxide hydrolases catalyze the hydrolysis reaction of the epoxide ring in alkene or arene compounds to form dihydrodiol.

Epoxide hydrolase-catalyzed reaction is represented in the chemical equation as shown below, where water is the co-substrate. The addition of water to an epoxide catalyzed by epoxide hydrolases produces 1,2-diols.

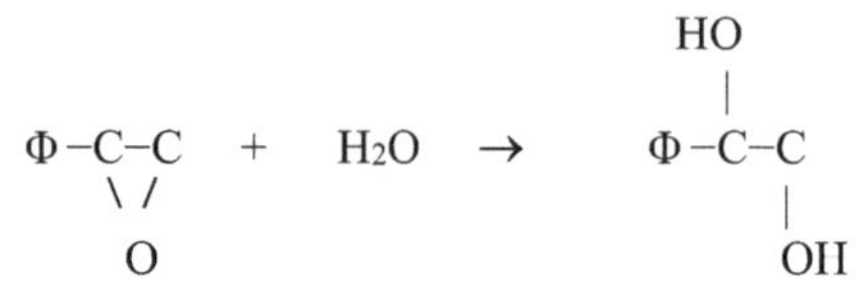

$$\Phi-\underset{\underset{O}{\diagdown\diagup}}{C-C} + H_2O \rightarrow \Phi-\underset{\underset{OH}{|}}{\overset{\overset{HO}{|}}{C}}-C$$

where Φ denotes an aromatic derivative. Other epoxide hydrolase-catalyzed reactions are such as the hydration of benzo(a)pyrene and allylbenzene oxide to form benzo(a) pyrene diol and allylbenzene diol, respectively.

8.3 Catalytic Actions on Atom or Molecule

Phase II detoxification enzyme-catalyzed reactions can occur at different atoms or molecules of foreign compounds, including phenol, epoxide, polyphenol, carboxylic acid, and amino acid. Specific atoms or molecules involved in conjugation reactions are discussed below, where R and Φ denote aliphatic and aromatic derivatives, respectively.

8.3.1 Conjugation at O Atom

Conjugation at O atom occurs in uridine-diphosphate-glucuronosyltransferase, sulfotransferase, or methyltransferase-catalyzed reactions. These reactions are described below:

(a) Uridine-Diphosphate-Glucuronosyltransferase-Catalyzed Reaction

Oxygen atom of hydroxyl group in phenol is conjugated with glucuronic acid (GA) to form a glucuronide derivative according to the reaction as shown below:

$$\Phi-OH \ + \ UDP-GA \ \rightarrow \ \Phi-O-GA \ + \ UDP$$

(b) Sulfotransferase-Catalyzed Reaction

Oxygen atom of hydroxyl group in alcohol or phenol is conjugated with sulfonate to form a sulfonate derivative, as shown in the following equation:

8.3 Catalytic Actions on Atom or Molecule

$$\Phi-OH + PAPS \rightarrow \Phi-O-SO_3H + PAP$$

where PAPS and PAP denote 3-phosphoadenosine 5- phosphosulfate and 3-phosphoadenosine 5- phosphate, respectively.

(c) Methyltransferase-Catalyzed Reaction

Oxygen atom of hydroxyl group in phenol is conjugated with methyl group of S - adenosylmethionine to form methyl conjugate as presented below:

$$\Phi-OH + SAM \rightarrow \Phi-O-CH_3 + SAH$$

In the equation, methyl group in SAM binds to O atom in phenol, where SAM and SAH denote S-adenosylmethionine and S-adenosyl-L-homocysteine, respectively.

8.3.2 Conjugation at N Atom

(a) Uridine-Diphosphate-Glucuronosyltransferase-Catalyzed Reaction

Glucuronic acid (GA) attaches to N atom of amine to form a glucuronide conjugate, as presented in the following reaction:

$$\Phi-\underset{H}{\overset{OH}{N}} + UDP\text{-}GA \rightarrow \Phi-\underset{GA}{\overset{OH}{N}} + UDP$$

(b) Sulfotransferase-Catalyzed Reaction

Sulfonate binds to N atom of aromatic amine to form sulfonate conjugate, as shown below:

$$\Phi-NH_2 + PAPS \rightarrow \Phi-\underset{H}{N}-SO_3H + PAP$$

(c) Methyltransferase-Catalyzed Reaction

Methyl group of SAM attaches to N-atom of amine to form methyl conjugate, as presented in the following equation:

$$\Phi\text{-NH-R} + \text{SAM} \rightarrow \Phi\text{-N-R} + \text{SAH}$$
$$\text{with } CH_3 \text{ on N}$$

$$\Phi\text{-NH-R} + \text{SAM} \rightarrow \overset{\underset{|}{CH_3}}{\Phi\text{-N-R}} + \text{SAH}$$

(d) Acetyltransferase-Catalyzed Reaction

Acetyl group of acetyl-CoA binds to N atom of amine to form acetyl conjugate as shown below:

$$\Phi-NH_2 + \text{Acetyl-CoA} \rightarrow \Phi-NH-COCH_3 + \text{CoA}$$

where Acetyl-CoA serves as the cofactor of the enzyme.

8.3.3 Conjugation at C Atom

In glutathione S-transferase (GST)-catalyzed reaction, reduced glutathione (GSH) binds to C atom of epoxide to form a glutathione conjugate as shown below:

$$\Phi\text{-}\underset{\underset{H}{|}}{\overset{\overset{O}{/\backslash}}{C}}\text{-}\underset{\underset{H}{|}}{C}\text{-R'} + \text{GSH} \rightarrow R\text{-}\underset{\underset{H}{|}}{\overset{\overset{OH}{|}}{C}}\text{-}\underset{\underset{SG}{|}}{C}\text{-R'}$$

8.3.4 Conjugation at S Atom

(a) Uridine-Diphosphate-Glucuronosyltransferase-Catalyzed Reaction

Glucuronic acid binds to the S atom of thio compound to form a glucuronide conjugate as shown below:

$$R-SH + UDP-GA \rightarrow R-S-GA + UDP$$

(b) Methyltransferase-Catalyzed Reaction

The methyl group attached to the sulfonium ion in SAM binds to the S atom of thio compound to form methyl conjugate, as shown below:

$$\Phi-SH + \text{SAM} \rightarrow \Phi-S-CH_3 + \text{SAH}$$

8.3.5 Conjugation of Carboxylic Acid

(a) Uridine-Diphosphate-Glucuronosyltransferase-Catalyzed Reaction

Glucuronic acid reacts with carboxylic acid to form a glucuronide conjugate, according to the following equation:

$$\Phi\text{-COOH} + \text{UDP-GA} \rightarrow \Phi\text{-}\underset{\underset{\text{OGA}}{|}}{\overset{\overset{O}{\|}}{C}} + \text{UDP}$$

(b) N-acyltransferase-Catalyzed Reaction

An amino acid reacts with a carboxylic acid group to form an amide, where Acyl-CoA serves as the cofactor for N-acyltransferase in the following conjugation reaction.

$$\Phi\text{-COOH} + \underset{\underset{R}{|}}{\text{HN-CH}_2\text{COOH}} \rightarrow \Phi\text{-CO-}\underset{\underset{R}{|}}{\text{N-CH}_2\text{COOH}} + \text{H}_2\text{O}$$

8.4 Other Non-conjugation Reactions

Other non-conjugation reactions include quinone reductase and epoxide hydrolase-atalyzed reactions. They are described below.

(a) Quinone reductase-Catalyzed Reaction

NADPH is involved in the reduction of quinone to hydroquinone according to the following reaction:

$$\underset{\underset{O}{\|}}{\overset{\overset{O}{\|}}{\Phi}} + 2\,\text{NADPH} \rightarrow \underset{\underset{\text{OOH}}{|}}{\overset{\overset{\text{OH}}{|}}{\Phi}} + 2\,\text{NADP}$$

(b) Epoxide Hydrolase-Catalyzed Reaction

Epoxide is hydrolyzed to form dihydrodiol, based on the equation below:

$$\Phi-\overset{O}{\overset{/\ \backslash}{C-C}}-R' \quad + \quad H_2O \quad \rightarrow \quad \Phi-\underset{OH}{\overset{OH}{\underset{|}{\overset{|}{C}}-C}}-R'$$

Bibliography

Almazroo OA, Miah MK, Raman Venkataramanan R (2017) Drug metabolism in the liver. Clin Liver Dis 21(1):1–20

Armstrong RN (1987) Enzyme-catalyzed detoxication reactions: mechanisms and stereochemistry. CRC Crit Rev Biochem 22:39–88

Armstrong RN (1991) Glutathione S-transferases: reaction mechanism, structure, and function. Chem Res Toxicol 4:131–140

Baez S, Segura-Aguilar J, Widersten M et al (1997) Glutathione transferases catalyse the detoxication of oxidized metabolites (o-quinones) of catecholamines and may serve as an antioxidant system preventing degenerative cellular processes. Biochem J 324:25–28

Berhane K, Widersten M, Engström A et al (1994) Detoxication of base propenals and other alpha, beta-unsaturated aldehyde products of radical reactions and lipid peroxidation by human glutathione transferases. Proc Natl Acad Sci USA 91:1480–1484

Butterworth M, Lau SS, Monks TJ (1996) 17 beta-Estradiol metabolism by hamster hepatic microsomes. Implications for the catechol-O-methyl transferase-mediated detoxication of catechol estrogens. Drug Metab Dispos 24:588–594

Capel ID, Millburn P, Williams RT (1974) The conjugation of 1- and 2-naphthols and other phenols in the cat and pig. Xenobiotica 4:601–615

Chasseaud LF (1979) The role of glutathione and glutathione S-transferases in the metabolism of chemical carcinogens and other electrophilic agents. Adv Cancer Res 29:175–274

Chen C-H (2012) Activation and detoxification enzymes: functions and implications. Spring Sciences, New York

Chen C-H (2020) Xenobiotic metabolic enzymes: bioactivation and antioxidant defense. Springer Nature, Cham

Ciotti M, Lakshmi VM, Basu N et al (1999) Glucuronidation of benzidine and its metabolites by cDNA-expressed human UDP glucuronosyltransferases and pH stability of glucuronides. Carcinogenesis 20:1963–1969

Coles B, Ketterer B (1990) The role of glutathione and glutathione transferases in chemical carcinogenesis. Crit Rev Biochem Mol Biol 25:47–70

Cook IT, Duniec-Dmuchowski Z, Kocarek TA et al (2009) 24-hydroxycholesterol sulfation by human cytosolic sulfotransferases: formation of monosulfates and disulfates, molecular modeling, sulfatase sensitivity, and inhibition of liver x receptor activation. Drug Metab Dispos 37:2069–2078

Duffel MW, Marshal AD, McPhie P et al (2001) Enzymatic aspects of the phenol (aryl) sulfotransferases. Drug Metab Rev 33:369–395

Fjellstedt TA, Allen RH, Duncan BK et al (1973) Enzymatic conjugation of epoxides with glutathione. J Biol Chem 248:3702–3707

Green MD, Tephly TR (1998) Glucuronidation of amine substrates by purified and expressed UDP-glucuronosyltransferase proteins. Drug Metab Dispos 26:860–867

Hammock BD, Storms DH, Grant DF (1997) Epoxide hydrolases. In: Guengerich FP (ed) Biotransformation. Elsevier Sciences, New York

Hayes JD, Judah DJ, McLellan LI et al (1991) Contribution of the glutathione S- transferases to the mechanisms of resistance to aflatoxin B1. Pharmacol Ther 50:443–472

Hempel N, Gamage N, Martin JL, McManus ME (2007) Human cytosolic sulfotransferase SULT1A1. Int J Biochem Cell Biol 39(4):685–689

Hirom PC, Idle JR, Millburn P et al (1977) Glutamine conjugation of phenylacetic acid in the ferret. Biochem Soc Trans 5:1033–1035

Hitchcock M, Smith JN (1964) Comparative detoxication. 13. Detoxication of aromatic acids in arachnids: arginine, glutamic acid and glutamine conjugations. Biochem J 93:392–400

Honma W, Shimada M, Sasano H et al (2002) Phenol sulfotransferase, ST1A3, as the main enzyme catalyzing sulfation of troglitazone in human liver. Drug Metab Dispos 30:944–949

Hutt AJ, Caldwell J (1990) Amino acid conjugation. In: Mulder GJ (ed) Conjugation reactions in drug metabolism. An integrated approach. Taylor and Francis, London

Ikenaka Y, Oguri M, Saengtienchai A et al (2013) Characterization of phase-II conjugation reaction of polycyclic aromatic hydrocarbons in fish species: unique pyrene metabolism and species specificity observed in fish species. Environ Toxicol Pharmacol 36(2):567–578

Ioannides C (2002) Enzyme systems that metabolise drugs and other xenobiotics. John Wiley, New York

Jernström B, Seidel A, Funk M et al (1992) Glutathione conjugation of trans-3,4-dihydroxy 1,2-epoxy 1,2,3,4-tetrahydrobenzo[c]phenanthrene isomers by human glutathione transferases. Carcinogenesis 13:1549–1555

Jin CJ, Miners JO, Burchell B et al (1993) The glucuronidation of hydroxylated metabolites of benzo[a]pyrene and 2-acetylaminofluorene by cDNA-expressed human UDP glucuronosyltransferases. Carcinogenesis 14:2637–2639

Keen JH, Jakoby WB (1978) Glutathione transferases. Catalysis of nucleophilic reactions of glutathione. J Biol Chem 253:5654–5657

Ketterer B (1986) Detoxication reactions of glutathione and glutathione transferases. Xenobiotica 16:957–973

Kiehlbauch CC, Lam YF, Ringer DP (1995) Homodimeric and heterodimeric aryl sulfotransferases catalyze the sulfuric acid esterification of N-hydroxy-2 acetylaminofluorene. J Biol Chem 270: 18941–18947

King CD, Rios GR, Green MD, Tephly TR (2000) UDP-glucuronosyltransferases. Curr Drug Metab 1:143–161

Klaassen CD, Boles JW (1997) Sulfation and sulfotransferases 5: the importance of 3′--phosphoadenosine 5′-phosphosulfate (PAPS) in the regulation of sulfation. FASEB J 11: 404–418

Kolm RH, Danielson UH, Zhang Y et al (1995) Isothiocyanates as substrates for human glutathione transferases: structure-activity studies. Biochem J 311:453–459

Lemke LE, McQueen CA (1995) Acetylation and its role in the mutagenicity of the antihypertensive agent hydralazine. Drug Metab Dispos 23:559–565

McGurk KA, Brierley CH, Burchell B (1998) Drug glucuronidation by human renal UDP-glucuronosyltransferases. Biochem Pharmacol 55:1005–1012

McLellan LI, Wolf CR, Hayes JD (1989) Human microsomal glutathione S-transferase. Its involvement in the conjugation of hexachlorobuta-1,3-diene with glutathione. Biochem J 258: 87–93

Minchin RF (1995) Acetylation of p-aminobenzoylglutamate, a folic acid catabolite, by recombinant human arylamine N-acetyltransferase and U937 cells. Biochem J 307:1–3

Minchin RF, Hanna PE, Dupret JM et al (2007) Arylamine N-acetyltransferase I. Int J Biochem Cell Biol 39:1999–2005

Mulder GJ (ed) (1990) Conjugation reactions in drug metabolism. An integrated approach. Taylor and Francis, London

Mulder GJ, Jakoby WB (1990) Sulfation. In: Mulder GJ (ed) Conjugation reactions in drug metabolism. An integrated approach. Taylor and Francis, London

Negishi M, Pedersen LG, Petrotchenko E et al (2001) Structure and function of sulfotransferases. Arch Biochem Biophys 390:149–157

Orzechowski A, Schrenk D, Bock-Hennig BS et al (1994) Glucuronidation of carcinogenic arylamines and their N-hydroxy derivatives by rat and human phenol UDP-glucuronosyltransferase of the UGT1 gene complex. Carcinogenesis 15:1549–1553

Parker MH, McCann DJ, Mangold JB (1994) Sulfation of di- and tricyclic phenols by rat liver aryl sulfotransferase isozymes. Arch Biochem Biophys 310:325–331

Ritter JK (2000) Roles of glucuronidation and UDP-glucuronosyltransferases in xenobiotic bioactivation reactions. Chem Biol Interact 129:171–193

Servin AL, Wicek D, Oryszczyn MP et al (1987) Metabolism of 6,7-dimethoxy 4-(4'-chlorobenzyl) isoquinoline. II. Role of liver catechol O-methyltransferase and glutathione. Xenobiotica 17:1381–1391

Temellini A, Mogavero S, Giulianotti PC et al (1993) Conjugation of benzoic acid with glycine in human liver and kidney: a study on the interindividual variability. Xenobiotica 23:1427–1433

Weber WW, Vatsis KP (1993) Individual variability in p-aminobenzoic acid N-acetylation by human N-acetyltransferase (NAT1) of peripheral blood. Pharmacogenetics 3:209–212

Weisiger RA, Pinkus LM, Jakoby WB (1980) Thiol S-methyltransferase: suggested role in detoxication of intestinal hydrogen sulfide. Biochem Pharmacol 29:2885–2887

Zamek-Gliszczynski MJ, Hoffmaster KA, Nezasa K-I et al (2006) Integration of hepatic drug transporters and phase II metabolizing enzymes: mechanisms of hepatic excretion of sulfate, glucuronide, and glutathione metabolites. Eur J Pharm Sci 27(5):447–486

Zeldin DC, Kobayashi J, Falck JR et al (1993) Regio- and enantiofacial selectivity of epoxyeicosatrienoic acid hydration by cytosolic epoxide hydrolase. J Biol Chem 268:6402–6407

Zeldin DC, Wei S, Falck JR et al (1995) Metabolism of epoxyeicosatrienoic acids by cytosolic epoxide hydrolase: substrate structural determinants of asymmetric catalysis. Arch Biochem Biophys 316:443–451

Reactive Intermediates and Their Interactions

9

Metabolism of foreign compounds aims to produce aqueous soluble metabolites so as to facilitate their excretion from the body. However, in many cases, metabolic conversion produces highly reactive metabolites. Such reactive intermediate formation during metabolic activation is an important mechanism attributing to foreign compound–mediated toxicity. Many toxic effects of foreign compounds do not result from the parent compounds, instead from reactive intermediates or metabolites that are formed during metabolism.

The toxicity of many foreign compounds is associated with their enzymatic conversion to toxic metabolites. Such metabolic process is commonly referred to as bioactivation. Metabolic activation is the consequence of metabolism of a foreign compound to reactive intermediate or metabolite. Accordingly, a toxic effect of a foreign compound is determined by not only the chemical nature of foreign compound but also the enzymatic reaction that metabolizes it.

9.1 Reactive Intermediate Species

The increased reactivity of metabolic intermediate or metabolite is primarily the result of conversion into electrophiles or free radicals. Many metabolic intermediates or metabolites are electrophiles, although some nucleophiles are reactive and many of them are further activated by conversion to electrophiles. There are nonionic and cationic electrophiles. Nonionic electrophiles include aldehydes, ketones, epoxides, quinones, sulfoxides, nitroso compounds, and acyl halides. While cationic electrophiles include carbonium ions and nitrenium ions.

As discussed previously, lipophilic foreign compounds are metabolized in two phases. In phase I, a functional group, such as hydroxyl or carboxyl, is introduced into a foreign compound by a phase I enzyme–catalyzed reaction. Subsequently, in phase II metabolism, an endogenous molecule, such as glucuronic, sulfuric, or amino acid, is added to the functionalized foreign compound by a phase II–catalyzed reaction.

© The Author(s), under exclusive license to Springer Nature Switzerland AG 2024
C.-H. Chen, *Activation and Detoxification Enzymes*,
https://doi.org/10.1007/978-3-031-55287-8_9

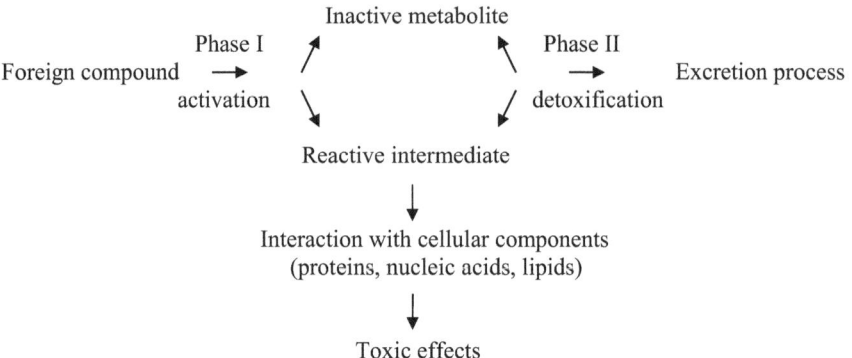

Fig. 9.1 Activation, detoxification, and toxic effects of a foreign compound

Hence, metabolic enzymes that are important in catalyzing bioactivation include not only phase I enzymes such as CYP450, flavin-containing monooxygenase, prostaglandin synthetase, and alcohol dehydrogenase but also phase II enzymes such as glutathione S-transferase, Uridine-diphosphate-glucuronosyltransferases, and sulfotransferase. A significant number of foreign compounds have potential to form reactive chemical intermediates, although functionalization reactions catalyzed by phase I enzymes convert some foreign compounds into unharmful substances.

Figure 9.1 illustrates reactive intermediate formation from foreign compound metabolisms. The figure reveals that phase I metabolism either yields inactive metabolites that are ready for phase II metabolism or produces highly chemically reactive intermediates that are capable of interacting with cellular components, if they are not quickly detoxified by phase II metabolism.

CYP450 is the most important enzyme that catalyzes the formation of electrophiles. Although some foreign compounds are oxidized by CYP450 isozymes to nontoxic metabolites, many CYP450 enzymatic reactions produce reactive intermediates during the oxidation of foreign compounds.

Reactive intermediates produced by phase I activation reactions are often electrophilic species, free radicals, or modified chemicals. Electrophiles include cations such as H^+ and NO^+; polarized neutral molecules such as HCl, alkyl halides, acyl halides, and carbonyl compounds; and polarizable neutral molecules such as Cl_2 and Br_2. Electrophile is a compound that is attracted to electrons and hence is a Lewis acid. Most electrophiles are positively charged, have an atom carrying a partial positive charge, or possess an atom not having an octet of electrons. An electrophile takes part in a chemical reaction that involves binding to a nucleophile.

While foreign compound metabolism serves as an important protective biological process, in some cases, it leads to producing adverse effects within the organism. When chemically reactive intermediates are created in sufficient quantities and are not promptly stabilized by endogenous substrates or other mechanisms, damage of cells and tissues can occur. Hence, it is essential to minimize the presence of reactive

9.1 Reactive Intermediate Species

intermediates or metabolites in intracellular levels and quickly remove them from the body.

9.1.1 Reactive Oxygen Species

Reactive intermediates can interact with oxygen molecules to produce reactive oxygen species. Reactive oxygen species are activated reactive intermediates, including highly reactive cationic electrophiles, such as carbonium, iminium, and nitrenium ions. Free radicals formed in metabolic activation include superoxide anion radical ($O_2^{-\bullet}$) and hydroxyl radical (HO•).

Reactive oxygen species are continuously produced during metabolic processes of foreign compounds. Meanwhile, they are detoxified by a variety of defense mechanisms. However, when the process of their generation is greater than that of detoxification, damages of cellular macromolecules occur.

Although a few reactive oxygen species may be implicated in cell signaling and in the immune system as a way to attack pathogens, highly reactive ions and free radicals are able to interact with cellular components (proteins, membrane lipids, and nucleic acids), resulting in damages to cellular molecules and their functions. For examples, peroxidases are a major enzyme group that catalyzes the formation of free radicals.

Figure 9.2 shows the generation of reactive oxygen species from reactive intermediates. It should be mentioned that the mechanism underlying free radical formation produced from reactive intermediates of foreign compounds is different from that generated from aerobic electron transfer chains in living organisms.

9.1.2 Reactive Nitrogen Species

Reactive nitrogen species (RNS) are various nitrogen derived compounds, such as nitroxyl anion, nitrosonium cation, nitrenium ion, and S-nitrosothiol. Similar to reactive oxygen species, elevated levels of RNS can also cause damages to cellular components, including proteins, lipids, and nucleic acids. At present, reactive nitrogen species generated by metabolic activation of foreign compounds is not as extensively investigated as that with reactive oxygen species.

Further investigations of reactive nitrogen species are needed, especially their effects on drug metabolisms. For instance, studies of the involvement of RNS in the therapeutic effects of acetaminophen have been carried out. Acetaminophen

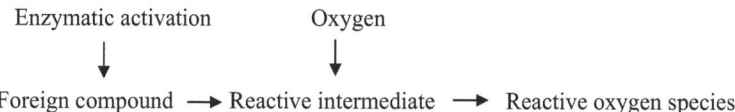

Fig. 9.2 Formation of reactive oxygen species

overdose was found to cause severe hepatotoxicity in humans as a result of the metabolism of acetaminophen to its reactive metabolite.

9.2 Enzyme-Catalyzed Reactive Intermediate Formation

CYP450 is the most important phase I activation enzyme with respect to the generation of reactive intermediates, such as epoxides, carbonium, iminium, and nitrenium ions, as well as free radicals. Other phase I enzymes, including peroxidase and flavin monooxygenases, also participate in catalytic reactions that produce reactive intermediates. In certain cases, conjugation reactions catalyzed by phase II enzymes are also involved. Such formation of reactive chemical intermediate mediated by metabolic enzymes are discussed below.

9.2.1 Mediation by Activation Enzymes

Some known foreign compounds and their reactive intermediate mediated by phase I activation enzymes are presented in Table 9.1, including drug ingredients such as aflatoxin and acetaminophen. Aflatoxin is a mycotoxin that is present in nature as aflatoxin B_1 produced by fungus Aspergillus. Crops are susceptible to infection by Aspergillus when they have prolonged exposure to a high humidity environment. While acetaminophen is one of the most commonly used medication as a pain and

Table 9.1 Reactive intermediates mediated by phase I catalytic reactions

Foreign compound	Activation enzyme	Reactive intermediate
Acetaminophen	CYP450	N-Acetyl-P-benzoquinoneimine
2-Acetylaminofluorene	CYP450	N-hydroxy-acetylaminofluorene
Acetylhydrazine	FMO	Acetyl radical or cation
Acrylaminde	CYP450	Glycidamide
Aflatoxin B1	CYP450/Peroxidase	Aflatoxin B1–8,9-epoxide
Benzo[a]pyrene	CYP450/epoxide hydrolase	Benzo[a]pyrene 7,8-dihydrodiol-9,10-epoxide
Bromobenzene	CYP450	Bromobenzene 3,4 oxide
4-Ipomeanol	CYP450	α,β-unsaturated dialdehyde
Menadione	CYP450	Semiquinone radical
2-Naphthylamine	CYP450	N-hydroxy-naphthylamine
N-Nitrosodimethylamine	CYP450	Carbonium ion
Phenylbutazone	PHS	Carbon-centered/peroxyl radical
Polycyclic aromatic compounds (PAH)	CYP/epoxide hydrolase	PAH diol-epoxides
Vinyl acetate	Carboxylesterase	Acetaldehyde

Abbreviations: *CYP450* as cytochrome P450, *PHS* as prostaglandine H synthase, and *FMO* as flavin monooxygenases

9.2 Enzyme-Catalyzed Reactive Intermediate Formation

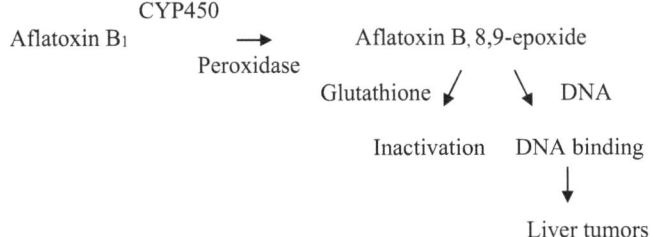

Fig. 9.3 Metabolic intermediate and toxic effect of aflatoxin B_1

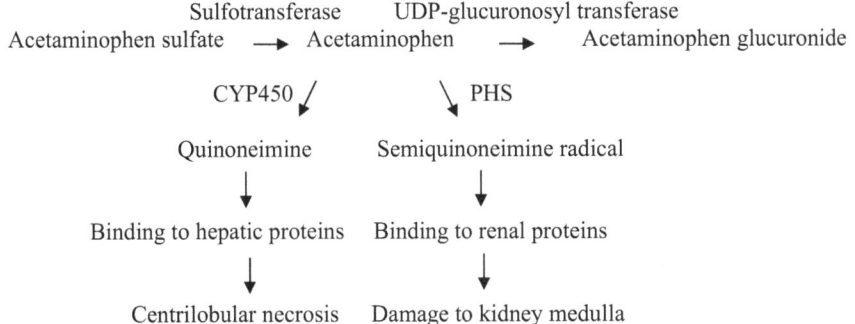

Fig. 9.4 Metabolic intermediates and toxic effects of acetaminophen

fever reliever. Its overdose-induced hepatotoxicity occurs mainly due to the production of N-acetyl-P- benzoquinoneimine.

While metabolic Intermediates and toxic effects of aflatoxin and acetaminophen are further shown in Figs. 9.3 and 9.4, respectively.

9.2.2 Mediation by Detoxification Enzymes

Metabolic reactive intermediates are mainly generated by phase I enzyme–catalyzed conjugation reactions. However, in certain cases, nonconjugation reactions catalyzed by phase II enzymes may also involve in their formation. For example, the formation of benzo[a]pyrene 7,8-dihydrodiol-9,10-epoxide, a benzo[a]pyrene intermediate, is catalyzed by a nonconjugation enzyme epoxide hydrolase.

Table 9.2 presents some examples of reactive intermediate formation mediated by phase II conjugation enzymes, such as UDP-glucuronosyltransferase, sulfotransferase, glutathione S-transferase, and N-acetyltransferase.

Table 9.2 Reactive intermediates mediated by phase II conjugation enzymes

Foreign compound	Conjugation enzyme	Reactive intermediate
2-Aminonaphthalene	UGT	N-hydroxy-2-naphthylamine
Benzidine	NAT	Acetyloxyarylamine nitrenium ion
Carboxylic acids	UGT	Acyl glucuronides
Dibromoethane	GST	Episulfonium ion
Dichloromethane	GST	Formaldehyde
Ethylene bromide	GST	1-bromo-2-S-glutathionyl ethane
Hexachlorobutadiene	GST	Mercapturic acid
7-hydromethyl,12-methyl-benz[a]anthracene	SULT	Carbonium ion
N-hydroxy-2-acetylaminofluorene	SULT	Nitrenium ion
N-Hydroxy-2-aminofluorene	NAT	Nitrenium ion
Tamoxifen	SULT	Tamoxifen carbocation

Abbreviations: *UGT* UDP-glucuronosyltransferases, *SULT* sulfotransferase, *GST* Glutathione S-transferases, and *NAT* N-acetyltransferases

9.3 Interactions with Cellular Components

When sufficient quantities are produced intracellularly, reactive intermediates including reactive oxygen and nitrogen species, have the capacity to interact with cellular components, including proteins, DNA and lipids, and ultimately to toxic effects. Such toxic effects may trigger alterations in the structure and function of target molecules. Consequently, reactive intermediates have been implicated in various disease conditions, such as allergic responses, and a variety of diseases, such as aging, cancer, cardiovascular disease, and neurological disorders.

The liver is not only the principal site for the activation of foreign compounds but also the center of reactive intermediate production as well as the primary site of toxic effects. Reactive chemical intermediates may also be exported to other tissues and cause harmful effects there, for example, aromatic amines are potential urinary bladder carcinogens.

When active metabolic intermediates are not effectively deactivated by nucleophilic species in the cells, covalent binding to proteins or DNA and lipid oxidation can occur. These can lead to protein and DNA adduct formation and lipid peroxidation. Such interactions of reactive intermediates with cellular components can be either specific or nonspecific. Specific interaction involves the binding to a specific site in a receptor, enzyme, or protein.

Figure 9.5 shows the interactions of reactive intermediates with proteins, DNA, or lipids, leading to the formation of protein adducts, DNA adducts, and lipid peroxidation, respectively. The figure presents the potential toxic effects as a result of interactions with cellular components.

9.3 Interactions with Cellular Components

```
                        Protein
Reactive intermediate   ────▶   Protein adduct   ────▶   Cytotoxicity

                        DNA
Reactive intermediate   ────▶   DNA adduct   ────▶   Mutation   ────▶   Malignancy

                        Lipid
Reactive oxygen species ────▶   Lipid peroxidation   ────▶   Tissue damages
```

Fig. 9.5 Interactions of reactive intermediates with cellular molecules

9.3.1 Protein Adducts

Primary binding sites of protein adducts are sulfur-containing amino acid residues cysteine and methionine, nitrogen of lysine, or histidine residues. Although not all metabolic intermediate bindings to protein are toxic in nature, protein adducts may be the source of toxicity caused by foreign compounds. The binding can also result in the inactivation of enzymatic functions, such as the binding of microcystin produced from blue-green algae to phosphatase.

9.3.2 DNA Adducts

The binding of reactive metabolic intermediate, particularly reactive oxygen species, to DNA base has received much attention. Among four bases in DNA that participate in the interaction, guanine residue is the predominant base for the binding. DNA adduct is not always stable. Hydrolysis can cleave the bond between purine and electrophilic metabolites. Moreover, DNA adduct formation can affect DNA base pairing. When DNA undergoes replication, erroneous base pairing can result in base mismatch. If base mismatch is not repaired during DNA replication, mutation can occur.

9.3.3 Lipid Peroxidation

Lipid peroxidation refers to oxidative degradation of membrane lipids. Cell damages occur when free radicals react with membrane lipids. Lipid peroxidation most often affects polyunsaturated fatty acids that contain multiple double bonds between methylene groups. The mechanism of lipid peroxidation consists of the following three major steps: initiation, propagation, and termination.

Initially, fatty acid free radical is produced when reactive oxygen species such as OH· reacts with unsaturated lipid. The resulting unstable fatty acid free radical can react with oxygen to yield peroxyl-fatty acid radical, which can then react with another fatty acid free radical to produce a different fatty acid radical and lipid peroxide. Such propagation process proceeds by a mechanism consisting of chain reactions.

To speed up the termination of the above-propagation process, living organisms have developed defense systems, such as antioxidant enzymes and small molecules that neutralize free radicals to protect living cells. However, if the produced free radicals are not terminated fast enough, damages to the cell components can occur. Consequently, the products of lipid peroxidation may be mutagenic and carcinogenic.

9.4 Factors Affecting Foreign Compound Toxicity

The amounts of metabolic intermediates that are present within the cells and their degree of toxicity that exerts on cellular components are dependent on relative rates of activation and detoxication reactions. While the metabolism of foreign compounds represents a protective biological process, an accumulation of metabolic intermediates within the cells may occur if the rate of the activation is higher than that of the detoxification. Hence, the balance between activation and detoxification processes can often determine whether a foreign compound causes toxic effects or not.

When metabolic intermediate or metabolite is responsible for the toxic effect of the parent compound, the toxicity of a foreign compound depends not only on the chemical nature of the parent compound but also on the enzymes that catalyze its metabolic reactions. Furthermore, foreign compound-metabolic enzymes are also affected by the genetic makeup and the environmental factors.

Environmental factors such as nutrition and chemical exposure are able to modulate the activities of metabolic enzymes. Hence, the extent of foreign compound–induced toxicity to an individual depends on the following several factors: (a) the chemical nature of the foreign compound, (b) the activity of its activation enzyme, (c) the activity of its detoxifying enzyme, (d) genetic polymorphisms of metabolic enzymes, (e) the environment that is exposed, and (f) lifestyle modifications.

9.5 Defense Against Reactive Intermediates

The protection against harmful effects of metabolic reactive intermediates, such as reactive oxygen and nitrogen species and free radicals, is to minimize their presence in intracellular levels. To achieve this goal, the body develops defense systems that include conjugation reactions, glutathione, and antioxidant enzymes. These defense systems are briefly described below.

9.5.1 Conjugation Reactions

In order to reduce the intracellular levels of reactive intermediates such as reactive oxygen and nitrogen species and free radicals, the body depends heavily on conjugation reactions catalyzed by phase II detoxification enzymes, including uridine-diphosphate-glucuronosyltransferases, glutathione S-transferases, and sulfotransferases.

Metabolic conjugation reaction usually represents a detoxication process. Conjugation reaction generally greatly increases aqueous solubility of foreign compound (with exception such as methylation), thus facilitating the excretion of foreign compound from the body. Extensive studies have shown that many conjugation reactions detoxify metabolites of foreign compounds. Nevertheless, conjugation of foreign compound with glucuronic acid, glutathione, or sulfonate in phase II metabolic reactions is thought to represent as a detoxification process for drugs and other chemicals; however, in some special cases, the resulting conjugates are harmful.

9.5.2 Glutathione

An electrophilic intermediate can interact with a nucleophilic species that contains an intramolecular center of high electron density. The body relies on such endogenous nucleophilic molecule to remove electrophilic intermediates. For example, glutathione, a nucleophilic species, is an important antioxidant for neutralizing free radicals produced by foreign compound metabolism. Glutathione reacts with chemically reactive intermediate to produce water-soluble compounds so as to facilitate the excretion of metabolites from the body.

9.5.3 Antioxidant Enzymes

The body also utilizes antioxidant enzymes as the primary line of defense in detoxifying metabolic reactive intermediates or free radicals that are produced in aerobic cellular metabolism. Among antioxidant enzymes present in the living organisms, superoxide dismutase converts the superoxide radical to form hydrogen peroxide and oxygen, and catalase and glutathione peroxidase work simultaneously with glutathione to reduce hydrogen peroxide to water and oxygen.

Besides serving as a primary line of defense in detoxifying free radicals produced in aerobic cellular metabolism, antioxidant enzymes can also effectively break down reactive oxygen species derived from metabolic conversion of foreign compounds. For examples, glutathione peroxidase catalyzes the reaction in which glutathione breaks down hydrogen peroxide and organic peroxide, while superoxide dismutase catalyzes the reaction that breaks down superoxide anions.

Moreover, the body also utilizes small antioxidant molecules to act as the primary line of defense in detoxifying free radicals produced in aerobic cellular metabolism. Such small antioxidant molecules, such as vitamins E and C and carotene, can neutralize free radicals by accepting or donating an electron, leading to the elimination of unpaired electron in free radicals. For instances, vitamin E protects cell membranes from oxidation damage by free radicals, and vitamin C scavenges free radicals and works with vitamin E to quench free radicals, and beta-carotene is an effective quencher of singlet oxygen.

Bibliography

Amacher DE (2006) Reactive intermediates and the pathogenesis of adverse drug reactions: the toxicology perspective. Curr Drug Metab 7:219–229

Anders MW (1985) Bioactivation of foreign compounds. Academic Press, New York

Anders MW (2007) Chemical toxicology of reactive intermediates formed by the glutathione-dependent bioactivation of halogen-containing compounds. Chem Res Toxicol 21:145–159

Baird WM, Hooven LA, Mahadevan B (2005) Carcinogenic polycyclic aromatic hydrocarbon-DNA adducts and mechanism of action. Environ Mol Mutagen 45:106–114

Boelsterli UA (2007) Mechanistic toxicology. CRC Press, Boca Raton

Bogdanffy MS, Taylor ML (1993) Kinetics of nasal carboxylesterase-mediated metabolism of vinyl acetate. Drug Metab Dispos 21:1107–1111

Chandrasekara A, Shahidi F (2011) Inhibitory activities of soluble and bound millet seed phenolics on free radicals and reactive oxygen species. J Agric Food Chem 59:428–436

Chen C-H (2012) Activation and detoxification enzymes: functions and implications. Springer Sciences, New York

Chen C-H (2020) Xenobiotic metabolic enzymes: bioactivation and antioxidant defense. Springer Nature, Cham

Dekant W, Vamvakas S (1993) Glutathione-dependent bioactivation of xenobiotics. Xenobiotica 23:873–887

Eaton DL, Gallagher EP (1994) Mechanisms of aflatoxin carcinogenesis. Annu Rev Pharmacol Toxicol 34:135–172

Glatt H (2000) Sulfotransferases in the bioactivation of xenobiotics. Chem Biol Interact 129:141–170

Guengerich FP (2001) Common and uncommon cytochrome P450 reactions related to metabolism and chemical toxicity. Chem Res Toxicol. 14:611–650

Hinson JA, Forkert PG (1995) Phase II enzymes and bioactivation. Can J Physiol Pharmacol 73:1407–1413

Jaeschke H, Knight TR, Mary Lynn Bajt ML (2003) The role of oxidant stress and reactive nitrogen species in acetaminophen hepatotoxicity. Toxicol Lett 144(3):279–288

James LP, Capparelli EV, Simpson PM et al (2008) Acetaminophen-associated hepatic injury: evaluation of acetaminophen protein adducts in children and adolescents with acetaminophen overdose. Clin Pharmacol Ther 84:684–690

Kalgutkar AS, Dalvie DK, O'Donnell JP et al (2002) On the diversity of oxidative bioactivation reactions on nitrogen-containing xenobiotics. Curr Drug Metab 3:379–424

Kim SY, Suzuki N, Laxmi YR et al (2004) Genotoxic mechanism of tamoxifen in developing endometrial cancer. Drug Metab Rev 36:199–218

Koob M, Dekant W (1991) Bioactivation of xenobiotics by formation of toxic glutathione conjugates. Chem Biol Interact 77:107–136

Levi PE, Hodgson E (2008) Reactive metabolites and toxicity. In: Smart RC, Hodgson E (eds) Molecular and biochemical toxicology. John Wiley, New York

Martínez MC, Andriantsitohaina R (2009) Reactive nitrogen species: molecular mechanisms and potential significance in health and disease. Antioxid Redox Signal 11(3):669–702

McLemore TL, Litterst CL, Coudert BP et al (1990) Metabolic activation of 4-ipomeanol in human lung, primary pulmonary carcinomas, and established human pulmonary carcinoma cell lines. J Natl Cancer Inst 82:1420–1426

Perlow RA, Kolbanovskii A, Hingerty BE et al (2002) DNA adducts from a tumorigenic metabolite of benzo[a]pyrene block human RNA polymerase II elongation in a sequence- and stereochemistry-dependent manner. J Mol Biol 321:29–47

Raucy JL, Kraner JC, Lasker JM (1993) Bioactivation of halogenated hydrocarbons by cytochrome P4502E1. Crit Rev Toxicol 23:1–20

Ritter JK (2000) Roles of glucuronidation and UDP-glucuronosyltransferases in xenobiotic bioactivation reactions. Chem Biol Interact 129:171–193

Shimada T (2006) Xenobiotic-metabolizing enzymes involved in activation and detoxificationof carcinogenic polycyclic aromatic hydrocarbons. Drug Metab Pharmacokinet 21:257–276

Smart RC, Hodgson E (2008) Molecular and biochemical toxicology. John Wiley, New York

Smith BJ, Curtis JF, Eling TE (1991) Bioactivation of xenobiotics by prostaglandin H synthase. Chem Biol Interact 79:245–264

Stadtman ER (2006) Protein oxidation and aging. Free Radic Res 40:1250–1258

Turrens JF (2003) Mitochondrial formation of reactive oxygen species. J Physiol 552:335–344

Metabolite-Associated Cell Toxicities 10

Potential toxic foreign compounds or metabolites are capable of interacting with an endogenous target, triggering perturbation in cellular functions and mediating a biochemical or biomedical effect. The cells initially respond to such perturbation with repair mechanisms. However, toxic effects occur as the induced perturbation exceeds the repair capacity. Then foreign compounds are absorbed and distributed to target organs, where they cause harmful effects. The liver and kidneys are frequent target organs of toxicity.

The exhibition of foreign compound toxicity at least occurs under the following various circumstances: (a) the presence of natural toxicity in a parent compound, such as toxigenic fungi; (b) the generation of reactive intermediates in activation metabolism, such as reactive oxygen and nitrogen species and free radical; and (c) the presence of induced toxicity by environment factors such as smoke and alcohol.

Biotransformation of the toxicities is one of the major factors that determine their presence and the duration at the site of effect. Toxicity is expressed as a function of two factors: dose and time as well as the type and intensity of the toxicity. These factors are directly dependent on the chemical transformation of the exposed parent substances. Toxicities including intrinsic, reactive intermediate–related, and lifestyle-induced toxicities are discussed below.

10.1 Intrinsic Toxicity

Toxigenic fungi are present in nature and also occur in food supplies due to mold infestation of susceptible agricultural products. When toxigenic fungi are present in sufficiently high levels, fungal metabolites exhibit toxic effects ranging from acute liver or kidney deterioration to chronic disease conditions such as liver cancer. Other foreign compounds that exhibit natural toxicity include nicotine, ethylene oxide, heavy-metal ions, strong acids or bases, and carbon monoxide.

Antioxidants such as indole-3-carbinol (I-3-C) show protection against hepatic toxicity, including carbon tetrachloride and other chemicals. Studies of intrinsic acute hepatic toxicity found that I-3-C produced modest increases in hepatic CYP450, UDP-glucuronosyl transferase, glutathione S-transferase, and NAD(P)H: oxidoreductase activities.

10.2 Reactive Intermediate–Related Toxicity

Reactive oxygen species, such as hydrogen peroxide, nitric oxide, hydroxyl radical, and superoxide anion radical, can directly damage cells or tissues. Reactive oxygen species form as chemically reactive intermediates react with an oxygen molecule. In addition, reactive nitrogen species also play an important role in xenobiotic-induced toxicity. Nitric oxide is generated enzymatically by nitric oxide synthase. If such generated reactive intermediates are not compensated by the body's antioxidant defense systems, the production of such reactive species prevails.

Detoxification reactions catalyzed by phase II enzymes serve as a defense system to detoxify such reactive species; however, an imbalance between activation and detoxification occurs if the rate of reactive intermediate species production is larger than that of detoxification reaction. As a result, oxidative stress occurs due to biological systems unable to readily detoxify such chemically reactive intermediates.

Oxidation stress is among the fundamental mechanisms as the results of the toxicity of many foreign compounds. The production of toxic intermediates occurs in metabolic activation of a foreign compound catalyzed by phase I enzyme systems, such as CYP450, flavin-monooxygenase, lipoxygenase, and xanthine oxidase.

Oxidative cell damages have been implicated in various disease conditions, such as neurodegeneration and cancer. Reactive oxygen and nitrogen species are also known to contribute as pathogenic factors to the development of such diseases at various stages.

Other than occurring during the processes of metabolic conversion, oxidative stress can also occur when the amount of reactive oxygen species produced in mitochondria through aerobic electron transfer chains is more than that needed for normal physiological functions, and also when the presence of cellular antioxidants cannot provide enough scavengers for these unwanted species.

10.3 Lifestyle-Induced Toxicity

Alcohol is hepatotoxic through metabolic disturbances associated with the oxidation of ethanol. Induction of microsomal enzymes also results in increased acetaldehyde generation from ethanol. Alcohol generated acetaldehyde promotes glutamyl-cysteinyl-glycine (GSH) depletion, free-radical-mediated toxicity, and lipid peroxidation. Ethanol is oxidized in liver microsomes, catalyzed by CYP450, resulting in ethanol tolerance and hepatic damages. CYP2E1 also activates various foreign compounds, leading to the increased susceptibility of the heavy drinker to toxicities.

Redox changes produced by nicotinamide adenine dinucleotide (NADH) generated via the liver pathway affect the metabolism of cellular components. Cigarette smoke is a well-established risk factor that contains toxic reactive molecules which are able to induce oxidative stress. Cigarette contributes to smoking-related diseases, including cardiovascular and pulmonary diseases. Cigarette smoke was also found to induce increases in protein carbonylation.

10.3.1 Alcohol

The enhanced toxicity of alcohol is a result of alcohol's effects on hepatocyte membranes, which renders the cells more susceptible to toxic injury. Alcohol activates CYP2E1 that catalyzes the activation of acetaminophen. Chronic ethanol consumption can potentiate acetaminophen hepatotoxicity through enhanced N-acetyl-p-benzoquinone imine formation.

Studies of the effect of chronic ethanol consumption have also revealed that acetaminophen toxicity as manifested by liver enlargement and congestion is significantly increased in the ethanol-treated objects.

10.3.2 Cigarette

Tobacco-specific nitrosamines are a group of carcinogens that are present in tobacco smoke. Reactive carbonyl species contained in cigarette smoke can markedly alter cell functions, such as proteins and lipids involved in metabolic processes. Cigarette smoking can also affect drug therapy associated with pharmacokinetic and pharmacodynamic mechanisms.

The nicotine-derived nitrosamines are strong carcinogens. Two tobacco-specific nitrosamine carcinogens are 4-(methylnitrosamino)-1-(3-pyridyl)-1-butanone (NNK) and N'-nitrosonornicotine (NNN). They are considered to be among the main causes of lung cancer in people who heavily use tobacco products. NNK and NNN are metabolized to diazonium ions and related electrophiles, which can react with DNA. A role for NNK in the induction of lung cancer by tobacco smoke is likely due to its specificity for the lung.

10.4 Toxic Effects on Cell Components

Major targets for oxidative damages are cellular molecules including proteins, DNA, and lipids. Xenobiotic-induced oxidative protein damage can cause an impairment of enzyme catalytic function. Oxidative DNA damage affects DNA base pairing, causing a mismatch in the DNA base pair transformation, resulting in a mutation. Oxidation of cell membranes can lead to peroxidation of a membrane lipid, a primary mechanism of tissue injury.

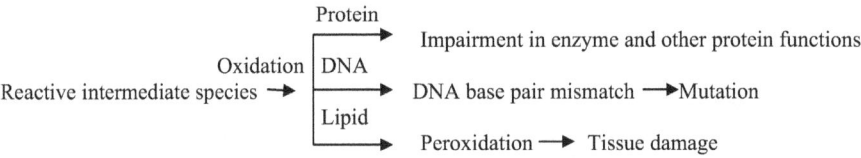

Fig. 10.1 Oxidative damages of cellular components by reactive metabolic intermediates

Reactive intermediates produced during foreign compound metabolism are believed to contribute to the development and progression of a variety of age-related diseases. For Instance, Acetaminophen is a primary ingredient in cold and pain relief remedies. In the liver, CYP450s are responsible for metabolism of acetaminophen. CYP450-catalyzed reaction metabolizes acetaminophen to produce reactive intermediates, quinoneimine, and semiquinoneimine, which can cause toxic effects by binding to hepatic and renal proteins. An overdose of drug acetaminophen can trigger potential liver damage, resulting in major acute liver failure.

Among industrial solvents, carbon tetrachloride undergoes activation metabolism catalyzed by CYP450 isozyme to form trichloromethyl radical. Such radical can trigger membrane lipid peroxidation. The liver and kidneys are major organs that are associated with such tetrachloride-induced toxicity.

Another example is 4-hydroxy-2-nonenal, which is a common product of polyunsaturated fatty acid oxidation. The decomposition of 4-hydroxy-2-nonenal can react with DNA and proteins, leading to depletion of intracellular glutathione, alteration of cell signaling cascades, and affecting multiple stress signaling pathways. The interactions of reactive metabolic intermediate species with cellular molecules are summarized in Fig. 10.1.

10.4.1 Protein Damage

Reactive oxygen species and other strong oxidants mediated by foreign compounds can interact with cellular proteins, resulting in oxidative protein damage. If such oxidative damages are not repaired, loss of catalytic function for enzymes and impairment of protein function can occur. Covalent binding of xenobiotic reactive metabolite to protein is associated with organ toxicity, such as acetaminophen, halothane, and 2,5-hexanedione, which form covalently bound adducts to protein targets.

Moreover, xenobiotic-induced oxidation of globin chains of hemoglobin can lead to disulfide bond formation, resulting in the interference of protein folding and the loss of protein function. Among amino acid residues, sulfur-containing amino acids (cysteine and methionine) are particularly susceptible to oxidation by reactive oxygen species.

10.4.2 DNA Damage

The oxidation of nucleic acids by reactive intermediate species is another consequence of oxidative stress in living cells. Reactive oxygen species are capable of inducing DNA damage, which eventually lead to cell transformation and tumor initiation. The oxidation of DNA base by reactive oxygen species can affect the consequences on DNA base pairing. For examples, the oxidation at the guanine residue can result in the insertion of a wrong base (adenine) in DNA. If the change in base pairing persists without repair during the next replication cycle, DNA damage can occur.

Reactive oxygen and nitrogen species are also associated with tumor promotion and progression. Oxidative DNA damage can subsequently lead to mutations and eventually tumors.

Metals are another group of foreign compounds that are associated with oxidative DNA damage. Nickel sulfides and other nickel compounds can give rise to the production of highly reactive hydroxyl radicals that can attack DNA bases and cause oxidative DNA damage. Nickel compound binding to DNA also poses an increased risk of respiratory cancer.

10.4.3 Lipid Peroxidation

When reactive metabolic species are generated in the membrane compartment of the cell, membrane lipids can be oxidized, leading to membrane peroxidation and oxidative lipid damages. The mechanism of lipid peroxidation involves the oxidation of fatty acyl carbon chains and the formation of lipid peroxyl radicals. Lipid peroxidation is a propagating chain reaction, which can result in damaging lipid bilayers, the backbone of biomembranes.

Reactive carbonyl species can even diffuse from the cell to attack targets far away from the site of formation as the propagators of oxidative stress to cause cellular and tissue damages.

Furthermore, the peroxidation reaction also produces reactive aldehyde, which can expand the toxicity of foreign compounds. Membrane lipid peroxidation has been implicated as one of the mechanisms that cause xenobiotic-induced tissue injury. The product of lipid peroxidation can cause DNA damage. Hence, the prevention of lipid peroxidation is an essential process in aerobic organisms.

Moreover, peroxidation of membrane phospholipid acyl chains generates chemically reactive intermediate carbonyl species, such as alpha, beta-unsaturated aldehydes, di-aldehydes, and keto-aldehydes. Molecular modifications induced by reactive carbonyl species play a causal role in the aging process. Most of such biological effects are due to the capacity of reactive carbonyl species to react with cellular constituents.

10.5 Toxic Effects on Cellular Functions

As described above, the conversion of chemicals into reactive intermediate species, such as electrophilic compounds or free radicals, can potentially alter the structure and function of cellular macromolecules. When protective defenses are overwhelmed by excess toxicant insult, the effects of reactive intermediate species can lead to deregulation of cell signaling pathways and dysfunction of cellular molecules.

Major foreign compound-mediated biomedical and biochemical effects include mitochondrial function intervention, interaction with ion transporters, enzymatic function interference, lipid peroxidation, and immune suppression and stimulation effects. Such xenobiotic-mediated toxic effects are briefly below.

10.5.1 Intervention with Mitochondria Functions

Mitochondria are crucial to the maintenance of many cellular functions, particularly energy utilization. The production of ATP is central to the supply of energy needed for biosynthesis and transport processes. Inhibition of mitochondrial functions subsequently causes a decrease in the production of cellular ATP, leading to an impairment of energy supply.

Through their interference with electron transport chains or by inhibiting ATP synthetase activity, foreign compounds can impair mitochondria functions. For example, pentachlorophenol is formed by metabolism of hexachlorobenzene. When absorbed into the body by occupational exposure, pentachlorophenol can cause toxic effects through its ability to uncouple mitochondrial oxidative phosphorylation. Moreover, feeding hexachlorobenzene for months was found to result in uncoupling of oxidative phosphorylation of liver mitochondria.

Mitochondrial dysfunction has also been implicated in the pathogenesis of valproic acid (an antiepileptic drug)-mediated hepatotoxicity. Acute treatment of rat hepatocytes with valproic acid was found to result in oxidative stress. While glutathione was found to provide protection to hepatocytes against mitochondrial damage by valproic acid. Moreover, rotenone is another foreign compound that is capable of intervening with mitochondria function. It is an inhibitor of oxidative phosphorylation, because of its blocking electron transfer in NADH-dependent oxidoreductase to prevent the utilization of NADH as a substrate.

In the presence of uncouples, ATP is not formed by mitochondrial ATP synthase, because the proton motive force across mitochondria membranes is dissipated. 2,4-dinitrophenol can uncouple the coupling of electron transport and phosphorylation in mitochondria so as to greatly increase metabolic rate. This chemical was applied to cause the rapid loss of body fat. However, concerns about its side effects, 2,4-dinitrophenol is discontinued for use in weight loss purposes.

10.5.2 Interaction with Ion Transporters

Membrane proteins have the capability of transporting specific ions across biological membranes. Channels and pumps are ion transporters. Interference their binding sites with a foreign compound may inhibit their ion transporter capacities. Na^+, K^+ pump is an ATP-dependent transporter that mediates the efflux of Na^+ in exchange of influx of K^+, leading to build up a Na^+ gradient across cell membranes. The generated Na^+ gradient can drive Na^+-dependent secondary transport.

For example, in cardiomyocytes, the influx of extracellular Na^+ into the cell drives the efflux of Ca^{2+} from the cell. The concentration of intracellular Ca^{2+} appears to associate with the contraction in cardiac muscle cells. Mediating the inhibition of Na^+, K^+ pump by binding to the outer surface of ATPase, digitoxin can block Na^+ and K^+ translocation as well as ATP hydrolysis. Specific inhibition of Na^+, K^+ pump by digitoxin indirectly causes an increase in the intracellular Ca^{2+} levels.

Such increase in intracellular Ca^{2+} concentration appears to directly stimulate contraction in cardiac muscle cells. Hence, digitalis is used in treating cardiac conditions such as congestive heart failure. However, an overdose of digitoxin can cause adverse drug reactions. Over-stimulation of cardiac muscle cells can lead to cardiac arrhythmia. Digitoxin overdose incidence occurs especially in aging populations.

Another example of foreign compound interaction with ion transporter is tetrodotoxin, a potent neurotoxin produced by marine bacteria. Poisoning with tetrodotoxin occurs after ingestion of various species of puffer fish. Tetrodotoxin specifically binds and inhibits a sodium channel in excitable nerve tissues. It blocks action potentials in nerves by binding to the pores of the voltage-dependent sodium channels in nerve cell membranes, leading to preventing the affected nerve cells from firing.

10.5.3 Interference with Enzymatic Functions

Interference with enzymatic function refers to the conditions when the enzyme cannot carry out its normal function as the catalyst for a specific reaction. Many enzymes and proteins contain a sulfhydryl group at their catalytic site. These enzymes are vulnerable to thio-reactive compounds, such as heavy metals that exhibit high affinity for the thiol group. For example, arsenic has been known to selectively inhibit pyruvate dehydrogenase. Pyruvate dehydrogenase is an enzyme complex which links the glycolytic pathway with the citrate cycle.

The binding of nonmetallic compounds at a different site may also cause enzyme inactivation. When inhibited or damaged by a toxic foreign compound, an enzyme may no longer carry out the conversion of the substrate. For example, acetylcholinesterase is a serine esterase whose function is to hydrolyze the neurotransmitter acetylcholine. Binding of organophosphate to acetylcholinesterase inhibits its enzyme activity. In the presence of organophosphate, the active serine hydroxyl group of acetylcholinesterase is attacked by phosphorus of the organophosphate, which results in the inhibition of acetylcholinesterase.

Enzymatic function can also be interfered by environment factors such as chronic consumption of alcohol. CYP2E1 is one pathway involved in oxidative stress produced by ethanol. CYP2E1 contributes significantly to ethanol metabolism and formation of acetaldehyde, the highly reactive metabolite. A number of potential toxic foreign compounds and their metabolites are able to weaken the body's antioxidant defense by interacting with antioxidant enzymes. Antioxidants are substances that either directly or indirectly protect cells against adverse effects of oxidants. Antioxidant enzymes provide scavengers for reactive oxygen species. For instance, superoxide dismutase, a typical antioxidant enzyme, scavenges a superoxide free radical by converting it to peroxide, which can then be destroyed by catalase.

10.5.4 Immune Suppression and Stimulation Effects

Repression of an immune cell development by toxic xenobiotics or metabolites may lead to immune suppression. Suppression of the immune system renders an organism susceptible to infections, compromising the ability of the body to fight diseases. Immunosuppressive agents include drugs that inhibit or reduce the activation or efficacy of the immune system. The side-effects of immunosuppressive agents include hypertension, peptic ulcers, and liver and kidney injury.

Moreover, immune suppressants can also cause harmful effects by interacting with other medicines. Typical examples of potential immune suppressants include steroids such as cortisone and environmental chemicals such as halogenated aromatic hydrocarbons. In addition, the environmental pollutant dioxin has a capacity of eliminating maturing T cells in the thymus.

Direct addition of alkylators to naive murine splenocytes has been reported to produce a dose-dependent suppression of the antibody-forming cell in response to antigens.

In contrast to immunosuppressive effect, some other foreign chemicals or metabolites can cause an enhancement in immune response, leading to tissue damage and immune-mediated disease. Stimulation of the immune system may result in hyposensitivity to immunoallergic reactions. A chemical allergy is initiated by the immune system and expressed as hypersensitivity. Enhancement of the immune system can also lead to chemical hypersensitivity. Such an adverse reaction to a chemical is the result of previous sensitization to that chemical or a structurally similar compound. After an initial allergic reaction to that chemical, a small subsequent exposure can evoke a much severe response. For instance, penicillin is the most common drug that can draw an allergic response to a minority of recipients.

10.6 Chemical Carcinogenesis

Chemical carcinogens belong to a diverse group of compounds. A significant number of chemicals that are carcinogenic act directly on the target cells. Others take effect after these chemicals are converted to toxic metabolites or intermediates through metabolic reactions catalyzed by phase I activation enzymes. Mutagenic chemicals are DNA damaging agents, which can alter genetic substances in the cells.

A variety of human carcinogens have been reported, including 2-naphthylamine, aflatoxins, arsenic and nickel compounds, asbestos, benzene, benzidine, tamoxifen, tobacco smoke, and polychlorinated biphenyl. Among DNA damaging agents, there are direct-acting carcinogens and indirect-acting carcinogens. Direct-acting carcinogens are compounds that are intrinsically reactive and can covalently interact with DNA to form DNA adduct. Typical direct-acting carcinogens include N-methyl-N-nitrosourea and methyl methanesulfonate.

While indirect-acting carcinogens require metabolic activation of parent compounds to produce reactive intermediates before they can covalently bind to DNA to form DNA adducts. In general, reactive intermediates produced during metabolic activation are electrophilic metabolites that are capable of interacting with nucleophilic sites in DNA. Such indirect-acting carcinogens include aflatoxin B1, dimethylnitrosamine, and benzo[a]pyrene.

10.7 Drug Metabolism Interference

Investigations of metabolic activation of chemicals to reactive intermediates is essential to the development and discovery of drugs. Reactive intermediates may bind to cellular macromolecules such as proteins and DNA, leading to oxidative stress. Hence, the concept of reactive intermediate formation during the biotransformation of drugs is an important bioactivation mechanism.

In addition, genetic factors may cause interindividual and intraindividual differences in drug metabolism. They may alter the balance between toxification and detoxification reactions. Besides, environmental factors such as alcohol and cigarette smoke can induce or inhibit drug-metabolizing enzymes and cause individual variation.

Most pharmacologically active molecules are lipophilic. Lipid-soluble xenobiotic is enzymatically transformed into polar, water-soluble metabolites to be excretable. The liver is the major organ for drug biotransformation. One of the major enzyme systems that determines the organism's capability of dealing with drugs is CYP450 monooxygenases. Other enzyme systems include dehydrogenases, oxidases, esterases, reductases, and a number of conjugating enzyme systems, such as glucuronosyltransferases, sulfotransferases, and glutathione S-transferases.

Metabolic bioactivation of drug molecules to form reactive metabolites is one of the mechanisms that can lead to hepatotoxicity or adverse drug reactions. Bioactivation of relatively inert functional groups to reactive metabolites may contribute towards certain drug-induced adverse reactions. Investigations into the role of xenobiotic-induced toxicity link between metabolic reactive intermediates and carcinogenesis. If reactive metabolites are not detoxified, they can covalently modify essential cellular targets. The resulting modification of DNA by reactive intermediates potentially leads to carcinogenesis. The formation of drug – protein adducts can also cause a potential risk of clinical toxicities.

Bibliography

Argikar UA, Mangold JB, Harriman SP (2011) Strategies and chemical design approaches to reduce the potential for formation of reactive metabolic species. Curr Top Med Chem 11(4):419–449

Arispe N, Diaz JC, Simakova O et al (2008) Heart failure drug digitoxin induces calcium uptake into cells by forming transmembrane calcium channels. Proc Natl Acad Sci 105:2610–2615

Blaikie FH, Brown SE, Samuelsson LM et al (2006) Targeting dinitrophenol to mitochondria: limitations to the development of a self-limiting mitochondrial protonophore. Biosci Rep 26:231–243

Brzezinski MR, Boutelet-Bochan H, Person RE et al (1999) Catalytic activity and quantitation of cytochrome P-450 2E1 in prenatal human brain. J Pharmacol Exp Ther 289:1648–1653

Chen C-H (2020) Xenobiotic metabolic enzymes: bioactivation and antioxidant defense. Spring Nature, Cham

Choi DW, Leininger-Muller B, Wellman M et al (2004) Cytochrome p-450-mediated differential oxidative modification of proteins: albumin, apolipoprotein E, and CYP2E1 as targets. J Toxicol Environ Health A 67:2061–2071

Chou AP, Li S, Fitzmaurice AG et al (2010) Mechanisms of rotenone-induced proteasome inhibition. Neurotoxicology 31:367–372

Colombo G, Garavaglia ML, Astori E et al (2019) Protein carbonylation in human bronchial epithelial cells exposed to cigarette smoke extract. Cell Biol Toxicol 35(4):345–360

Eickhorn R, Weirich J, Hornung D, Antoni H (1990) Use dependence of sodium current inhibition by tetrodotoxin in rat cardiac muscle: influence of channel state. Pflugers Arch 416:398–405

Goetz ME, Luch A (2008) Reactive species: a cell damaging rout assisting to chemical carcinogens. Cancer Lett 266:73–83

Gonzalez FJ (2005) Role of cytochromes P450 in chemical toxicity and oxidative stress: studies with CYP2E1. Mutat Res 569:101–110

Haggerty HG, Kim BS, Holsapple MP (1990) Characterization of the effects of direct alkylators on in vitro immune responses. Mutat Res 242:67–78

Hecht SS, Stepanov I, Steven G, Carmella SG (2016) Exposure and metabolic activation biomarkers of carcinogenic tobacco-specific nitrosamines. Acc Chem Res 49(1):106–114

Hodgon E, Smart RC (2001) Introduction to biochemical toxicology. Johnn Wiley, New York

Jaeschke H, Gores GJ, Cederbaum AI et al (2002) Mechanisms of hepatotoxicity. Toxicol Sci 65:166–176

Kalabus JL, Cheng Q, Jamil RG et al (2012) Induction of carbonyl reductase 1 (CBR1) expression in human lung tissues and lung cancer cells by the cigarette smoke constituent benzo[a]pyrene. Toxicol Lett 211(3):266–273

Kalgutkar AS, Soglia JR (2005) Minimising the potential for metabolic activation in drug discovery. Expert Opin Drug Metab Toxicol 1(1):91–142

Leung L, Kalgutkar AS, Obach RS (2012) Metabolic activation in drug-induced liver injury. Drug Metab Rev 44(1):18–33

Lieber CS (1994) Susceptibility to alcohol-related liver injury. Alcohol Alcohol Suppl 2:315–326

Luís PB, Ruiter JP, Aires CC et al (2007) Valproic acid metabolites inhibit dihydrolipoyl dehydrogenase activity leading to impaired 2-oxoglutarate-driven oxidative phosphorylation. Biochim Biophys Acta 1767:1126–1133

Masini A, Ceccarelli-Stanzani D et al (1985) The role of pentachlorophenol in causing mitochondrial derangement in hexachlorobenzene induced experimental porphyria. Biochem Pharmacol 34:1171–1174

Mates JM (2000) Effects of antioxidant enzymes in the molecular control of reactive oxygen species toxicology. Toxicology 153:83–104

Meyer UA (1996) Overview of enzymes of drug metabolism. J Pharmacokinet Biopharm 24(5): 449–459

Orhan H, Karakuş F, Ergüç A (2021) Mitochondrial biotransformation of drugs and other xenobiotics. Curr Drug Metab 22(8):657–669

Pamplona R (2008) Membrane phospholipids, lipoxidative damage and molecular integrity: a causal role in aging and longevity. Biochim Biophys Acta 1777:1249–1262

Pérez MJ, Cederbaum AI (2003) Adenovirus-mediated expression of Cu/Zn- or Mn-superoxide dismutase protects against CYP2E1-dependent toxicity. Hepatology 38:1146–1158

Port JL, Yamaguchi K, Du B et al (2004) Tobacco smoke induces CYP1B1 in the aerodigestive tract. Carcinogenesis 25(11):2275–2281

Shertzer HG, Sainsbury M (1991) Intrinsic acute toxicity and hepatic enzyme inducing properties of the chemoprotectants indole-3-carbinol and 5,10-dihydroindeno[1,2-b]indole in mice. Food Chem Toxicol 29(4):237–242

Spracklin DK, Hankins DC, Fisher JM et al (1997) Cytochrome P450 2E1 is the principal catalyst of human oxidative halothane metabolism in vitro. J Pharmacol Exp Ther 281:400–411

Tong V, Teng XW, Chang TK, Abbott FS (2005) Valproic acid II: effects on oxidative stress, mitochondrial membrane potential, and cytotoxicity in glutathione-depleted rat hepatocytes. Toxicol Sci 86:436–443

Wells PG, Kim PM, Laposa RR et al (1997) Oxidative damage in chemical teratogenesis. Mutat Res 396:65–78

West JD, Marnett LJ (2005) Alterations in gene expression induced by the lipid peroxidation product, 4-hydroxy-2-nonenal. Chem Res Toxicol 18:1642–1653

Oxidative Stress: Reactive Chemical Intermediates

Oxidative stress is characterized by an imbalance between the production of reactive chemical intermediates and the ability of organisms to neutralize such reactive chemical species, including reactive oxygen and nitrogen species, free radicals, and electrophiles. Highly reactive molecules can damage cells to cause disease conditions and aging related issues. Organs rely on detoxification metabolisms and antioxidants to carry out reactions to detoxify such reactive chemical intermediates.

Reactive oxygen species are highly reactive chemicals formed from oxygen, such as peroxides. Free radicals derived from cell metabolisms are highly reactive and unstable molecules that have unpaired electron, such as hydroxyl radicals and superoxide. While electrophiles are often positively charged molecules that do not have an octet of electrons. They are electron deficient species, such as hydronium ion, which can accept an electron pair from electron rich species.

Increasing evidence points to oxidative stress and inflammation as common pathological mechanisms underlying many disease conditions. Inflammation is a biological response to oxidative stress. Oxidative stress and chronic inflammation play a significant role in the initiation of carcinogenesis. The inflammation triggered by oxidative stress is the cause of many chronic diseases, such as neurodegeneration, cardiovascular disease, diabetes, and cancer.

Oxidative stress can activate a variety of transcription factors, leading to the differential expression of genes involved in inflammatory pathways. Reducing oxidative stress and chronic inflammation require an elevation of the levels of antioxidant enzymes and detoxification enzymes. Nrf2 is a transcription factor that induces the expression of cytoprotective and detoxification genes, thus reducing oxidative stress and inflammation. The Nrf2-ARE pathway is an intrinsic mechanism of defense against oxidative stress. Many evidences demonstrate the protective role of the Nrf2-ARE pathway in degenerative conditions.

11.1 Oxidative Stress

Oxidative stress occurs as a result of imbalance between the production of reactive chemical species derived from foreign compound metabolism and the ability of organisms to neutralize them. Reactive chemical species are such as peroxides and free radicals. High production of reactive chemical species is associated with a significant decrease in antioxidant defense mechanisms.

Interaction of reactive chemical species with cell organisms can cause damages to cell components, such as protein adducts, lipid peroxidation, and nucleic acid mutagenicity, which subsequently disrupt cellular functions. Oxidative stress and inflammation are common pathological mechanisms associated with major stages of cytotoxicity and carcinogenesis.

The cellular damages resulting from oxidative stress are a fundamental pathological process in a variety of diseases. Oxidative stress has been reported to play a role in the pathogenesis of neurodegenerative diseases. Aging is the main risk factor of neurodegeneration and chronic neuroinflammation is a pathological process underlying neurodegeneration.

Nuclear erythroid-2-related factor 2 (Nrf2) is a transcription factor that induces the expression of cytoprotective and detoxification genes, thus reducing oxidative stress and inflammation. The Nrf2-ARE (antioxidant response element) pathway is an intrinsic mechanism of defense against oxidative stress. Many evidences demonstrate the protective role of Nrf2-ARE pathway in various disease conditions caused by oxidative stress and inflammation.

11.1.1 Reactive Oxygen Species

Enzyme-catalyzed electron chains in mitochondria generate a small amount of reactive oxygen species for normal physiological functions. While the production of reactive oxygen species also occurs in xenobiotic metabolic processes that involve metabolic enzyme systems.

Reactive oxygen species are formed in the reaction of metabolic reactive intermediate with oxygen. Oxidative stress occurs as the produced reactive oxygen species are more than needed for normal physiological functions.

Metabolism of foreign compounds can also lead to reactive chemical intermediates, such as reactive oxygen species. While metabolic enzymes also play central roles in the detoxification of foreign compounds. However, when detoxification metabolism and cellular antioxidants cannot provide enough scavengers, the excessive reactive oxygen species are capable of oxidizing cellular molecules to cause oxidative cell damages.

11.1.2 Free Radicals

Free-radical metabolites can be generated during metabolism by reductase-catalyzed reaction or lipid peroxidation. Peroxidases are major enzymes that catalyze the formation of free radicals. Free radicals formed in metabolic activation include superoxide anion radical ($O_2^{\cdot-}$) and hydroxyl radical (HO•). Some free radicals can also react with oxygen to give superoxide anions.

Free radical species can bind covalently to cellular macromolecules and also can promote lipid peroxidation in cellular membranes. Such high reactivity of free radical can lead to an increase of oxidative stress. For examples, aging is considered as the result of oxidative damage to living cells. Metabolic activation of carbon tetrachloride to free-radical intermediates can lead to the acute liver injury.

Biochemical studies on the metabolic activation showed that halogenated alkanes and tetrachloride are metabolized to reactive free radicals. Free radical formation in the liver is associated with hepatotoxicity during administration of halogenated anesthetics. Oxidative stress induces a cellular redox imbalance, which has been found to be present in various cancer cells in comparison with normal cells.

11.2 Reactive Intermediate–Mediated Oxidative Stress

Foreign compound–induced oxidative stress includes (a) interaction with cellular proteins, leading to a loss of enzyme catalytic function; (b) oxidation of cell membranes, resulting in peroxidation of membrane lipid; (c) interaction with DNA, thereby affecting DNA base pairing; and (d) mismatch in DNA base pair, leading to mutation.

Figure 11.1 illustrates oxidative stress as a result of an imbalance between the production of foreign compound metabolic reactive intermediates and the body's defense mechanism.

11.2.1 Oxidative Stress on Biomolecules

Cellular molecules are major targets for oxidative damages. Foreign compounds–mediated oxidative protein damage can cause an impairment of enzyme catalytic functions or other protein functions. Oxidative DNA damage can affect DNA base pairing, causing a mismatch in the DNA base pair transformation and resulting in a mutation. Oxidation of cell membranes can lead to peroxidation of membrane lipid. Figure 11.2 illustrates the interactions of chemically reactive intermediates with proteins, lipids, and DNA, which lead to the formation of protein adducts, DNA adducts, and lipid peroxidation.

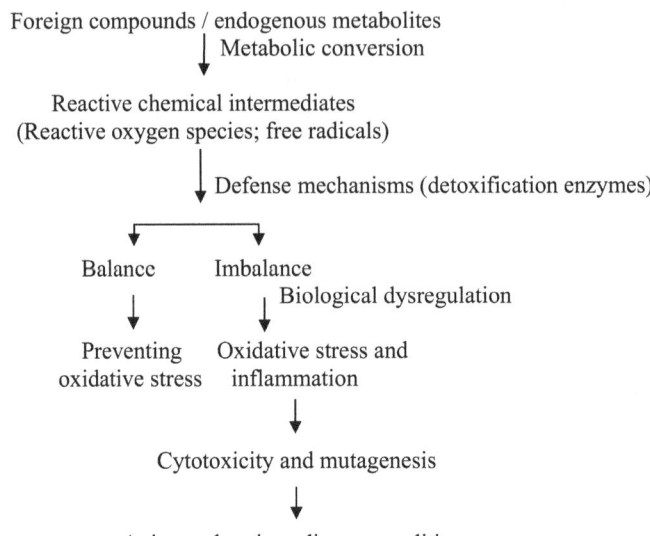

Fig. 11.1 Oxidative stress and biological dysregulation

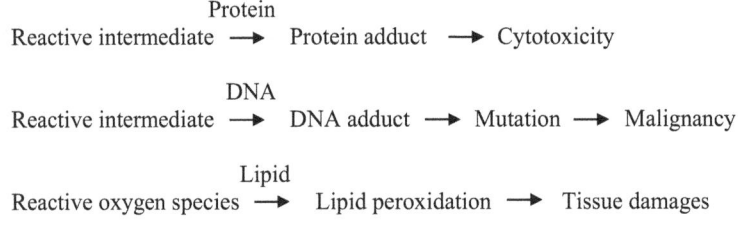

Fig. 11.2 Interactions of reactive intermediates with cellular molecules

(a) Protein Functions

Reactive intermediates mediated by foreign compounds are able to interact with cellular proteins, resulting in oxidative protein damage. Oxidative stress damage can affect protein folding, leading to the loss of enzyme catalytic function and impair of various protein functions. Among amino acid residues, the sulfur-containing amino acids, cysteine and methionine, are particularly susceptible to oxidation by reactive oxygen species. For example, hemoglobin is a frequent target for oxidative damage. Xenobiotics-induced oxidation of two globin chains of hemoglobin can lead to disulfide bond formation, resulting in the interference of its folding and the loss of its function.

Besides sulfur-containing amino acid residues, the primary binding sites of protein adducts include nitrogen of lysine or histidine residues. Although not all metabolic intermediate bindings to protein are toxic in nature, protein adducts are the source of toxicity mediated by foreign compounds. The binding can result in inactivation of enzymatic functions, such as in the case of microcystin binding to phosphatase, which leads to the inactivation of the enzyme.

Channels and pumps are membrane proteins that have the capability of transporting specific ions across biological membranes. Interference of foreign compounds with channel or pump at its binding site may inhibit the capacity of ion transporter. For example, tetrodotoxin, a potent neurotoxin, specifically binds and inhibits sodium channel in excitable nerve tissues. Another example is digitoxin, which mediates an inhibition of sodium or potassium pump by binding to its specific amino acid residues.

(b) Lipid Peroxidation

Reactive oxygen species are able to oxidize cell membrane lipids, resulting in the peroxidation of membrane lipids and causing damage to cellular membranes. Peroxidation of membrane lipids is a propagating chain reaction that can also jeopardize lipid bilayers.

The mechanism of lipid peroxidation involves the oxidation of fatty acyl carbon chains and the formation of lipid peroxyl radical. The reaction also produces reactive aldehyde, which can expand the toxicity of xenobiotics.

A typical example of membrane lipid peroxidation is triggered by carbon tetrachloride. Carbon tetrachloride is an industrial solvent that undergoes bioactivation metabolism catalyzed by CYP450 isomers to form trichloromethyl radical. Major organs that are associated with tetrachloride-induced toxicity are the liver and kidney.

(c) DNA Damages

The oxidation of DNA bases by reactive oxygen species has severe consequences on DNA base pairing. If the change in base pairing persists without repair during the next replication cycle, DNA damage can occur, such as in the case of DNA base oxidation at guanine residue, which results in the insertion of a wrong adenine base. DNA damage can also give rise to the production of highly reactive hydroxyl radicals that can attack DNA bases. Oxidative DNA damage can subsequently lead to mutation and eventually tumor.

11.2.2 Oxidative Stress and Inflammation on Diseases

Inflammation is the body's response to a variety of oxidative stress. Oxidative stress–mediated chronic inflammation is a major cause of age-related diseases. Inflammation induces a reduction in cellular antioxidant capacity, leading to uncontrolled inflammatory response that can give rise to cell and tissue destruction.

Growing evidences have suggested that elevated level of oxidative stress plays a crucial role in various pathogenesis processes contributing to human diseases, such as chronic inflammation, diabetes, neurodegeneration, cardiovascular disorders, and cancer.

11.3 Electrophilic Stress

Electrophiles are electron deficient species. They accept an electron pair from electron rich species, such as carbocations and carbonyl compounds. They are often positively charged or molecules that do not have an octet of electrons. Electrophiles react by accepting an electron pair in order to form a bond to a nucleophile. In opposite, a nucleophile is an electron rich species, which donates electron pairs to electron deficient species.

Humans are exposed to xenobiotic electrophiles through the environment, lifestyle, and dietary habits. The majority of cellular electrophiles are generated from polyunsaturated fatty acids by a peroxidation chain reaction. Reactive electrophile species levels need to be controlled in healthy cells. Their formation and destruction during electrophilic stress is of great concern. Moreover, reactive electrophilic species can potentially affect gene expression at various levels by chemically reacting with nucleic acids and proteins. Although excess reactive electrophilic species production can lead to cell damage, lower levels of reactive electrophilic species may modulate the expression of cell survival genes and may actually contribute to survival during severe stress.

11.3.1 Foreign Compound–Mediated Electrophiles

Foreign compounds may be transformed to form harmful products by metabolic activation, which are often referred to as metabolic reactive intermediates. They are metabolized in two phases. In phase I activation mechanism, a functional group, such as hydroxyl or carboxyl, is introduced into a foreign compound. Followed by phase II detoxification metabolism, in which an endogenous molecule, such as glucuronic, sulfuric, or amino acid, is added to the functional group of the foreign compound.

Formation of highly reactive metabolites from relatively inert or nontoxic chemical compounds is referred to bioactivation. Its related increase in chemical reactivity is primarily the result of conversion into electrophiles or free radicals. The formation of electrophiles from xenobiotics accounts for most toxicities, which is referred to as foreign compound–mediated electrophilic stress.

Electrophiles can be inherently reactive or be generated by metabolism of chemical compounds to reactive intermediates. An electrophile derived from xenobiotics can react with a nucleophile on their molecules to covalently modify them and cause toxicity. For instance, reaction with proteins can lead to cellular toxicity and immunogenicity, and reaction with nucleic acids can lead to gene mutation and carcinogenesis. Xenobiotic electrophiles appear to act as signaling molecules and activation of various redox signaling pathways is involved in cell proliferation, detoxification, or excretion of electrophiles.

11.3.2 Varieties of Electrophiles

Exogenous reactive chemicals can impair cellular homeostasis and are often associated with the development of various disease conditions. Significant progress has been made by studying the macromolecular interactions with chemicals that possess various electron-withdrawing groups, as well as the elucidation of the protective responses of cells to chemical interventions. Elimination of toxic reactive intermediates and metabolites as well as prevention of their formation involve detoxication mechanisms.

Electrophilic metabolites consist of nonionic and cationic electrophiles. Nonionic electrophiles include aldehydes, ketones, epoxides, quinones, sulfoxides, nitroso compounds, and acyl halides. Cationic electrophiles include carbonium ions and nitrenium ions. Electrophilic metabolites are mainly detoxicated by enzyme-catalyzed oxidation and hydrolysis reactions, such as in the cases of epoxide, arene, and organophosphate. Other electrophiles are such as Ag^+, Cd^{2+}, and Hg^{2+}, which can be carried out by conjugation with glutathione.

A variety of phase I enzymes, including CYP450 and peroxidases, bioactivate nitrogen-containing foreign compounds via direct oxidations on the nitrogen atom. The reaction leads to forming reactive intermediates, such as the N-oxidation of aromatic amines. Moreover, peroxides are also generated in cells that consume oxygen. Among different peroxides, hydrogen peroxide is the molecule that is formed in the highest quantities.

Bioactive metabolites often interact with the body tissues to precipitate toxicities. The bioactivation reactions are catalyzed by CYP450 and other phase I activation enzymes. For example, polycyclic aromatic hydrocarbon naphthalene is bioactivated by CYP450 to an electrophilic epoxide intermediate and is subsequently metabolized to naphthoquinones and to a free radical intermediate. These reactive intermediates may bind covalently to tissues and cause oxidant stress or lipid peroxidation.

11.3.3 Electrophilic Stress on Biomolecules

As discussed above, reactive oxygen species are able to interact with cell components, including proteins, nucleic acids, and lipids. Electrophiles derived from foreign compounds can react with nucleophiles on macromolecules to covalently modify them. Reaction with proteins can lead to cellular toxicity and immunogenicity. Reaction with lipids can cause lipid oxidation, While reaction with nucleic acids can lead to gene mutation and carcinogenesis. Their bioactive metabolites often interact with the body tissues to mediate carcinogenesis.

For examples, α,β polyunsaturated lipid aldehydes are potent lipid electrophiles that covalently modify lipids, proteins, and nucleic acids. Protein carbonylation, a major outcome of oxidative stress, is implicated as an initiating factor in mitochondrial dysfunction and endoplasmic reticulum stress. Moreover, reactive oxygen species can directly cause covalent modifications to DNA. They can also initiate the formation of lipid hydroperoxides.

The consequence of lipid hydroperoxide-mediated DNA damage in cardiovascular diseases is an important area of research. The body's antioxidant systems include antioxidant enzymes and small antioxidant molecules. Antioxidants provide scavengers for reactive oxygen species and serve as the brake to prevent their over presence.

11.3.4 Electrophiles on Disease Conditions

Oxidative stress caused by the imbalance between reactive oxygen species production and detoxification has been implicated in many disease conditions, such as cardiovascular risk, hypertension, metabolic disorder, and cancer. Oxidative stress–mediated inflammation and cellular damages are believed to be linked to cardiovascular diseases. By interfering with electron transport chain or inhibiting ATP synthetase activity, mitochondria functions may be impaired by xenobiotic metabolic reactive intermediates.

Reactive oxygen species can also promote carcinogenesis in cells. Xenobiotic-induced mitochondrial toxicity can result in a decrease in cellular ATP production, leading to the impairment of energy supply and interference with mitochondrial functions. Studies have established that oxidative stress is a significant threat to the integrity of neurons. Many central nervous system disorders points toward oxidative damages of neurons. The level of charged species and free radicals is related to oxidative damages to neuronal cells.

Moreover, repression of immune cell development by xenobiotics may lead to immune suppression, as in the case of environmental pollutant such as dioxin. Suppression of the immune system can make the organism more susceptible to infections. Chronic inflammation, such as atherosclerosis disease of the vascular system, is the leading cause of cardiovascular diseases.

Excessive generation of reactive oxygen species leads to a state of oxidative stress, which is a major risk factor for the development and progression of atherosclerosis. However, hydrogen peroxide not only exerts essential physiological functions but also is a universal toxin that organisms deploy to kill invading cells. While pentachlorophenol and valproic acid are capable of inhibiting electron transport chain by binding to an electron transport chain component.

11.3.5 Electrophiles on Drug Metabolism

The metabolic activation of a drug to an electrophilic reactive metabolite and its covalent binding to cellular macromolecules are associated with the occurrence of idiosyncratic drug toxicity. Drug-induced liver injury can be caused by several mechanisms. Metabolic bioactivation of drugs to form reactive metabolites is considered as an initiation mechanism.

The formation of reactive metabolites leads to oxidative and electrophilic stress. Drug-induced liver injury is a significant leading cause of hepatic dysfunction. For example, acetaminophen is metabolized in the liver to N-acetyl-p-benzoquinone imine, an electrophilic metabolite known to bind liver proteins, resulting in hepatotoxicity.

11.4 Defense Against Oxidative Stress

Strategies for chemoprevention include decreasing metabolic phase I enzyme activity responsible for generating reactive metabolic species, as well as increasing phase II enzyme activity, which deactivates radicals and electrophiles known to intercede in cellular processes. Accordingly, a major mechanism of chemical protection against carcinogenesis, mutagenesis, and other toxicities mediated is the modification of metabolic enzymes, particularly phase II enzymes.

Accordingly, a hypothesis to explain the mechanisms of the chemoprevention activity is to activate detoxification systems such as glutathione S-transferase. A number of other enzymes, including glutathione peroxidase or reductase, catalase, and superoxide dismutase, are also important in the protection of cells from chemical-induced oxidative damages. For example, catalase and glutathione systems are applied to prevent cellular accumulation of peroxides and the damages generated by peroxide-derived radical.

11.4.1 Chemoprevention Inducers

Animal studies indicate that induction of phase II enzymes is a sufficient condition for obtaining chemoprevention. Fruits and vegetables or their natural constituents, which enhance detoxication enzyme–catalyzed metabolism, are considered as good candidates to prevent reactive chemical intermediate–induced toxicities. Such chemoprevention can be achieved by inducer agents such as 1,2-dithiole-3-thiones, terpenoids, and the isothiocyanate sulforaphane.

For examples, sulfur-containing substances derived from garlic and onion have been shown to prevent carcinogenesis. The effects of S-allylcysteine (SAC), a water-soluble organosulfur compound derived from garlic, on glutathione S-transferase (GST) activities in the liver were investigated to demonstrate that the administration of SAC significantly increases the level of GST in the liver. These findings support the hypothesis that SAC exhibits chemoprevention activity by exerting its effects on phase II detoxification enzymes.

Moreover, oltipraz was also considered as a potential chemoprevention agent. The chemoprevention efficacy achieved by administration of intermittent doses of oltipraz was evaluated in rat livers. Its chemoprevention actions were found to associate with the increases in the levels of phase II detoxifying enzymes, such as GST isozymes. Besides, the chemoprevention effects of organosulfur compounds S-methylcysteine were also examined. The results support S-methylcysteine and cysteine as chemoprevention agents for hepatocarcinogenesis.

11.4.2 Nrf2- ARE Pathway

Nrf2 is a transcription factor that regulates the cellular defense against toxic and oxidative insults through the expression of genes involved in oxidative stress response. Nrf2 complex binds to antioxidant response element (ARE) to promote the transcription of cytoprotective genes. The Nrf2-ARE pathway is involved in many cellular processes, including metabolism and inflammation.

By involving in the expressions of many detoxification enzymes and antioxidants, Nrf2 acts as a sensor of oxidative stress to prevent genome instability. The Nrf2-ARE pathway appears to be as a master regulator to protect cells from oxidative and electrophilic stresses. The subject of defense against oxidative stress by the Nrf2-ARE pathway is further discussed in another chapter.

Bibliography

Albano E, Lott KA, Slater TF et al (1982) Spin-trapping studies on the free-radical products formed by metabolic activation of carbon tetrachloride in rat liver microsomal fractions isolated hepatocytes and in vivo in the rat. Biochem J 204(2):593–603

Batty M, Bennett MR, Yu E (2022) The role of oxidative stress in atherosclerosis. Cell 11(23):3843

Cheeseman KH, Albano EF, Tomasi A, Slater TF (1985) Biochemical studies on the metabolic activation of halogenated alkanes. Environ Health Perspect 64:85–101

Chen C-H (2012) Activation and detoxification enzymes: functions and implications. Springer Science, New York

Chen C-H (2020) Xenobiotic metabolic enzymes: bioactivation and antioxidant defense. Springer Nature, Cham

Cuendet M, Oteham CP, Moon RC et al (2006) Quinone reductase induction as a biomarker for cancer chemoprevention. J Nat Prod 69(3):460–463

Dringen R, Pawlowski PG, Hirrlinger J (2005) Peroxide detoxification by brain cells. J Neurosci Res 79(1–2):157–165

Farmer EE, Davoine C (2007) Reactive electrophile species. Curr Opin Plant Biol 10(4):380–386

Franzoni F, Scarfo G, Guidotti S et al (2021) Oxidative stress and cognitive decline: the neuroprotective role of natural antioxidants. Front Neurosci 15:729757

Fukushima S, Takada N, Wanibuchi H et al (2001) Suppression of chemical carcinogenesis by water-soluble organosulfur compounds. J Nutr 131(3s):1049S–1053S

Goetz ME, Luch A (2008) Reactive species: a cell damaging rout assisting to chemical carcinogens. Cancer Lett 266:73–83

Halliwell B (2007) Oxidative stress and cancer: have we moved forward? Biochem J 401(1):1–11
Hatono S, Jimenez A, Wargovich MJ (1996) Chemopreventive effect of S-allylcysteine and its relationship to the detoxification enzyme glutathione S-transferase. Carcinogenesis 17(5):1041–1044
Hauck AK, David A, Bernlohr DA (2016) Oxidative stress and lipotoxicity. J Lipid Res 57(11):1976–1986
Hodgon E, Smart RC (2001) Introduction to biochemical toxicology. John Wiley, New York
Incalza MA, D'Oria R, Natalicchio A et al (2018) Oxidative stress and reactive oxygen species in endothelial dysfunction associated with cardiovascular and metabolic diseases. Vascul Pharmacol 100:1–19
Jaeschke H, Gores GJ, Cederbaum AI et al (2002) Mechanisms of hepatotoxicity. Toxicol Sci 65:166–176
Kalgutkar AS, Dalvie DK, O'Donnell JP et al (2002) On the diversity of oxidative bioactivation reactions on nitrogen-containing xenobiotics. Curr Drug Metab 3(4):379–424
Lee SH, Blair IA (2001) Oxidative DNA damage and cardiovascular disease. Trends Cardiovasc Med 11(3–4):148–155
Liebler DC, Guengerich FP (2005) Elucidating mechanisms of drug-induced toxicity. Nat Rev Drug Discov 4(5):410–420
Liguori I, Russo G et al (2018) Oxidative stress, aging and diseases. Clin Interv Aging 13:757–772
Mates JM (2000) Effects of antioxidant enzymes in the molecular control of reactive oxygen species toxicology. Toxicology 153:83–104
Pamplona R (2008) Membrane phospholipids, lipoxidative damage and molecular integrity: a causal role in aging and longevity. Biochim Biophys Acta 1777:1249–1262
Plummer JL, Beckwith AL, Bastin FN et al (1982) Free radical formation in vivo and hepatotoxicity due to anesthesia with halothane. Anesthesiology 57(3):160–166
Primiano T, Egner PA, Sutter TR et al (1995) Intermittent dosing with oltipraz: relationship between chemoprevention of aflatoxin-induced tumorigenesis and induction of glutathione S-transferases. Cancer Res 55(19):4319–4324
Qu Q, Liu J, Zhou HH, Klaassen CD (2014) Nrf2 protects against furosemide-induced hepatotoxicity. Toxicology 324:35–42
Sajadimajd S, Khazaei M (2018) Oxidative stress and cancer: the role of Nrf2. Curr Cancer Drug Targets 18(6):538–557
Sies H, Berndt C, Dean P, Jones DP (2017) Oxidative stress. Annu Rev Biochem 86:715–748
Singh E, Devasahayam G (2020) Neurodegeneration by oxidative stress: a review on prospective use of small molecules for neuroprotection. Mol Biol Rep 47(4):3133–3140
Thompson RA, Isin EM, Ogese MO et al (2016) Reactive metabolites: current and emerging risk and hazard assessments. Chem Res Toxicol 29(4):505–533
Vassort G, Turan B (2010) Protective role of antioxidants in diabetes-induced cardiac dysfunction. Cardiovasc Toxicol 10(2):73–86

Metabolic Enzymes: Polymorphisms and Species Differences

12

Before elimination from the body, foreign compounds to which humans are exposed undergo metabolic conversion mediated by phase I and phase II metabolic enzymes. Phase I activation enzymes catalyze oxidation, hydrolysis, and reduction reactions to metabolize foreign compounds. The produced functionalized compounds then proceed with reactions catalyzed by phase II detoxification enzymes to facilitate the elimination of water-soluble metabolites from the organism.

Extensive investigations have revealed that these foreign compound-metabolic enzymes exhibit genetic polymorphisms with two or more variants differing slightly in their amino acid sequences. Such enzyme polymorphism can give rise to the differences in the ability of individuals to metabolize foreign compounds.

Individual susceptibility to environments, chemicals, and drugs is also affected by polymorphisms in metabolic enzymes. The variant that differs slightly in the amino acid sequence may affect substrate specificity, enzymatic activity and related transport functions.

Genetic polymorphism is a DNA sequence variation. DNA sequence refers to the precise order of nucleotide bases that encode amino acids.

The metabolism of drugs and other foreign compounds into more hydrophilic metabolites is essential to facilitate their excretion from the body. Enzyme polymorphisms can influence drug efficacy and safety through alternating their pharmacokinetics and disposition of drugs. Such genetic differences in drug metabolism are the result of variations in alleles for genes that code for the enzymes responsible for drug metabolism. Individual differences in genes for drug metabolism are also important for diagnosing individuals who are prone to side effects from the drugs.

Besides, the expression of foreign compound-metabolic enzymes can also differ as a result of exposure to enzyme inducers. Both genetic polymorphisms and enzyme inducibility are of major significance, since these two factors play an important role in the susceptibility of individuals to xenobiotic toxicity. Genetic polymorphism

© The Author(s), under exclusive license to Springer Nature Switzerland AG 2024
C.-H. Chen, *Activation and Detoxification Enzymes*,
https://doi.org/10.1007/978-3-031-55287-8_12

may also act as molecular biomarkers to provide important predictive information about individual's response to cytotoxicity and carcinogenesis.

Many carcinogens show species differences in their activities. Since most carcinogens need to be metabolized to reactive electrophiles in order to elicit their toxic effects, the observed species differences in response to toxic effects have significance with respect to researching metabolism of foreign compounds. The subject of inducibility of metabolic enzymes is discussed in another chapter.

12.1 Enzyme Polymorphisms on Xenobiotic Metabolism

Genetic polymorphisms may lead to altering individual's response to cytotoxicity and carcinogenesis of foreign compounds. Reduction in metabolic capacity of activation enzymes can lead to lower metabolic rate. While reduction in metabolic capacity of detoxification enzymes can result in increased toxic effects of reactive intermediate species derived from xenobiotic metabolism.

Dietary or environmental toxins can bind to DNA, resulting in DNA mutation and leading to carcinogenesis. As a result, the interrelationship between genetic susceptibility and toxic exposure is important in the development of various disease conditions. Consequently, individual predisposing factors due to genetic polymorphisms in metabolic enzymes have been implicated in the molecular pathogenesis and the heterogeneity of various disease conditions.

Genetic variations in the levels of expression and substrate specificity of metabolic enzymes can also give rise to abnormal foreign compound metabolism. These variations may account for the differential susceptibility of individuals to potentially toxic xenobiotics. Genetic polymorphisms and molecular markers can be applied to expand clinical prevention studies targeted to individuals with a high risk of developing certain disease conditions.

12.2 Genetic Polymorphisms of Activation Enzymes

Genetic variability in the expression of activation enzymes is an important relevant feature associated with foreign compound–mediated biochemical and biomedical effects. Such differences in individual sensitivity to xenobiotic metabolites could explain many observed differential reactions to foreign compounds among individuals and the susceptibility of individual to diverse toxicities.

Figure 12.1 illustrates the effects of genetic polymorphisms of activation enzymes on foreign compound–modulated toxicity. A higher activation capacity can significantly increase metabolic rate. However, a much higher activation rate may lead to over production of reactive intermediates or metabolites.

Variants in the gene encoding of activation enzymes can result in altering enzymatic activity and bioactivation of foreign compounds. Genetic polymorphisms of major activation enzymes, such as cytochrome P450 oxidase, flavin containing monooxygenase, peroxidase, carboxylesterase, alcohol and aldehyde

12.2 Genetic Polymorphisms of Activation Enzymes

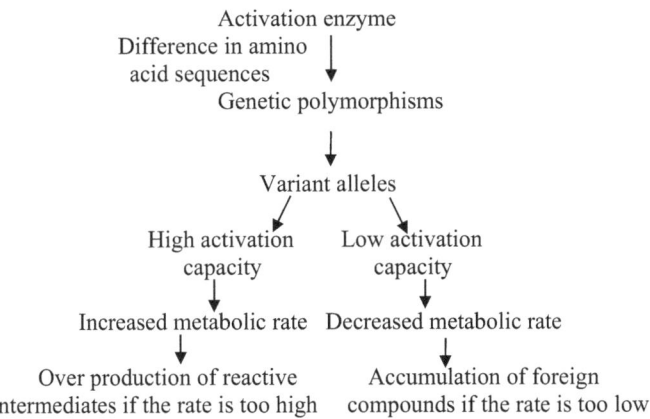

Fig. 12.1 Effects of activation enzyme polymorphisms on foreign compound metabolism

Table 12.1 Polymorphisms of activation enzymes

Activation enzymes	Genotype
Cytochrome P450 oxidase	CYP2E1, CYP2D6, CYP1A1
Flavin-containing monooxygenase	FMO3 E158K, FMO3 E308G
Peroxidase	MPO-463A
Carboxylesterase	CES1, CES2
Alcohol dehydrogenase	ADH1B
Aldehyde dehydrogenase	ALDH2 (ADH2*1, ADH2*2 and ADH2*3)

dehydrogenases, and their associated variations in activation metabolism, are summarized in Table 12.1.

Among phase I metabolic enzymes, cytochromes P450 (CYP450) and alcohol dehydrogenase (ALDH) are two major membrane-bound enzymes that catalyze activation reactions of xenobiotic compounds. Discussions of polymorphisms of CYP450 and ALDH enzymes and their association with toxicological actions are presented below.

12.2.1 Cytochrome P450 (CYP450) Polymorphisms

CYP450 enzyme polymorphisms exhibit significant differences in their substrate and product selectivity. Between individuals, there are also significant differences in the levels of CYP450 isozyme expressions.

Approximately 40% of human CYP450-dependent drug metabolism is carried out by polymorphic enzymes. Genetic polymorphisms including CYP1A1, 2A5, 2A6, 2E1, and 3A4 have been investigated to examine their impacts on individual susceptibility to foreign compound toxicities. These studies are discussed below.

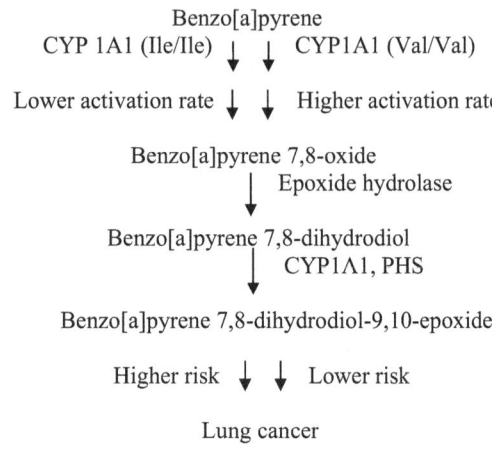

Fig. 12.2 CYP1A1 polymorphisms affecting benzo[a]pyrene activation

(a) CYP1A1

Hepatotoxicity is one of the most common toxicities observed in medication usages. The activation of CYP1A1 inhibition–mediated aryl hydrocarbon receptor was found to be involved in drug-induced hepatotoxicity. Research into the inhibitors of CYP1A1 has been carried out to improve chemoprevention.

CYP450 genetic polymorphisms play an important role in the differences between individuals in response to the toxicity of polycyclic aromatic hydrocarbons. Polymorphic variants in CYP450 gene, especially CYP1A1, have an effect on the metabolism of polycyclic aromatic hydrocarbons, thereby influencing the susceptibility of individuals towards their toxicities.

Figure 12.2 presents CYP1A1 polymorphisms affecting benzo[a]pyrene activation. Metabolism of aromatic compounds, such as benzo[a]pyrene in tobacco smoke, involves the activation of epoxide. The activated epoxide is capable of covalently binding to DNA. Thus, a higher level of DNA adducts is detected in lymphocytes of tobacco smokers with specific polymorphism of CYP1A1 gene. Moreover, specific genotype Val/Val for CYP1A1 was found to be related to the incidence of esophageal carcinoma, especially in heavy smokers.

(b) CYP2A6

High expression of CYP450 activity results in an undesired increase in the rate of activation of foreign compounds, which can lead to over-production of reactive metabolites.

Hence, metabolisms mediated by high activity of CYP450 have been implicated in the production of toxicity and carcinogenicity of certain foreign compounds. Polymorphic variants in CYP450 genes may lead to variation in the extent of such activation, thereby influencing individual sensitivity towards xenobiotic toxicity.

12.2 Genetic Polymorphisms of Activation Enzymes

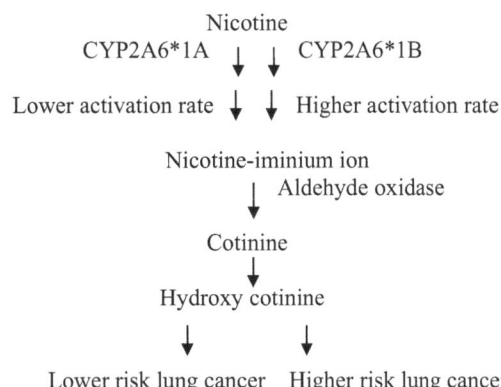

Fig. 12.3 Polymorphisms in CYP2A6 affect nicotine metabolism

CYP2A6 is the principle human nicotine C-oxidase involving the metabolic activation of nicotine. About 70–80% of nicotine is oxidized by CYP450 to nicotine-iminium ions, which are further metabolized to cotinine by aldehyde oxidase. Nicotine polymorphisms have been reported to influence the rate of nicotine elimination from the body. Interindividual and interethnic variabilities in the level of CYP2A6 activity are attributed to polymorphisms in the CYP2A6 gene.

Figure 12.3 presents CYP2A6 polymorphisms affecting nicotine metabolism. The produced nicotine-iminium ion can react with DNA, leading to higher lung cancer risk. CYP2A6 is also responsible for the clearance of many drugs and environmental chemicals.

The enzyme is present predominantly in the liver. Over 22 different alleles have been characterized for this enzyme. It also activates a number of structurally unrelated chemical carcinogens, including aflatoxin B1, halothane, and valproic acid.

(c) CYP2E1

CYP2E1, another polymorphic variant in CYP450 genes, is a key enzyme in the metabolic activation of a variety of toxicants, including benzene, vinyl chloride, and halogenated solvents. CYP2E1 is also one of the enzymes that metabolize ethanol to acetaldehyde. Genetic polymorphisms in CYP2E1 have been identified and linked to altered susceptibility of individuals to alcoholism. Moreover, CYP2E1 has many important physiological functions and is a key enzyme for hepatic carcinogenesis, drug toxicity, and liver disease.

As a contributing factor for alcohol-induced oxidative liver injury, CYP2E1 is also involved in the metabolism of nitrosamines in tobacco smoke. Moreover, genetic polymorphisms of CYP2E1 may lead to interindividual differences in CYP2E1- mediated drug oxidation activities, including the oxidation of N-nitrosamines.

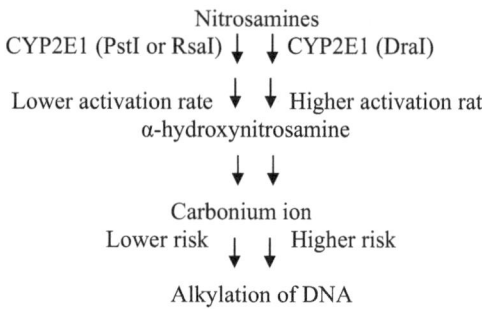

Fig. 12.4 Polymorphisms in CYP2E1 affect nitrosamine activation

Figure 12.4 illustrates polymorphic variants in CYP2E1 that affect metabolism of nitrosamines. The formed reactive intermediate, carbonium ion, can react with DNA. Studies have also demonstrated an association of the CYP2E1 gene with the susceptibility of individuals to lung cancer.

(d) Other CYP450 Polymorphisms

Other polymorphisms of CYP450 enzyme include CYP2A5, CYP2A6, and CYP3A4.

Among them, enzyme CYP2A5 can be induced by ethanol. Chronic feeding of ethanol to mice was found to increase CYP2A5 catalytic activity. Antioxidants can lower the alcohol elevation of reactive oxygen species and block the alcohol induction of CYP2A5.

The polymorphism of CYP2A6 also has a role in nicotine metabolism, since it catalyzes the metabolism of nicotine and the tobacco-specific lung carcinogen. Genetic variation in CYP2A6 may also affect smoking behavior and contribute to lung cancer risk.

Enzyme CYP3A4 is essential for the metabolism of many medications, such as the metabolism of many drugs. Variants in the CYP3A4 gene polymorphisms have their distinct functional significances in drug metabolism.

12.2.2 Flavin-Containing Monooxygenase

The majority of flavin-containing monooxygenase (FMO)-mediated metabolism in the liver is catalyzed by flavin-containing monooxygenase 3 (FMO3). FMO3 is a hepatic microsomal enzyme that oxidizes foreign compounds, such as drugs and other chemicals.

Numerous variants in the gene encoding FMO3 have been identified. Some of which result in altered enzymatic activity and substrate metabolism. Two commonly occurring polymorphisms of FMO3 are E158K (glutamic acid is replaced with lysine at position 159) and E308G (glutamic acid is replaced with glycine at position 308). They have been found to associate with a reduction in patients with adenomatous polyposis.

12.2.3 Peroxidase

Myeloperoxidase (MPO), an enzyme highly expressed in leukocytes, transforms benzo(a)pyrene and aromatic amines to reactive intermediates. MPO463A (alanine at position 463) polymorphism strongly affects MPO mRNA expression, suggesting that MPO463A variant may be a protective factor in the etiology of lung cancer.

12.2.4 Carboxylesterase

Carboxylesterase1 (CES1) and carboxylesterase2 (CES2) polymorphisms are responsible for the hydrolysis of ester and amide bonds present in pharmaceutical products. Genetic variants of CES1 and CES2 genes have been shown to influence drug metabolism. For example, polymorphisms of the CES1 gene have been reported to affect the metabolism of methylphenidate, a stimulant medication, while variants of the CES2 gene have been found to affect the metabolism of aspirin.

12.2.5 Alcohol and Aldehyde Dehydrogenases Polymorphisms

Alcoholism is a typical example that genetic polymorphisms affect the expression of metabolic enzymes. Alcohol dehydrogenase (ADH) is the enzyme that catalyzes the oxidation of alcohol to aldehyde. While aldehyde dehydrogenase (ALDH) oxidizes aldehyde to acetate, which is further metabolized to CO_2 and H_2O. The kinetics of ethanol absorption and elimination are influenced by genetic factors. ADH and ALDH variations in the genes and the environment contribute to such alcohol disorders.

ADH is present as a polygene family, which has seven genes, including ADH2*1, ADH2*2, and ADH2*3, which are functionally polymorphic. ALDH also exhibits genetic polymorphisms, including ALDH2*1 and ALDH2*2 alleles. Genetic variations in ADH and ALDH among individuals in response to alcohol have been reported to contribute to differences in susceptibility to alcohol abuse and dependence. Particularly, the functional genetic variants of ADH, which exhibit high alcohol oxidizing activity.

The genetic variants of ALDH that exhibit low acetaldehyde oxidizing activity protect against heavy drinking and alcoholism. A functional variant in alcohol dehydrogenase 1B (ADH1B) or aldehyde dehydrogenase 2 (ALDH2) is protective in most of people.

Figure 12.5 illustrates ADH and ALDH polymorphisms that affect alcohol metabolism.

While the effects of enzyme polymorphisms on alcoholism and smoker are discussed in Sect. 12.4.

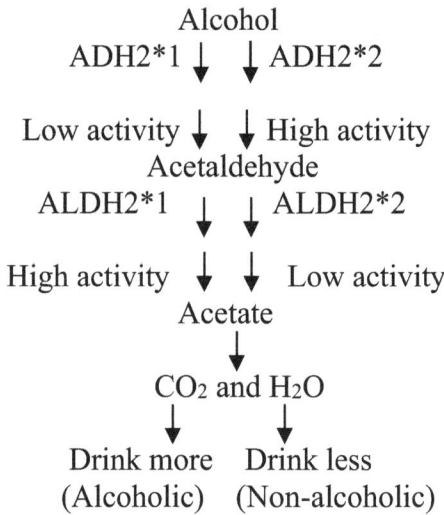

Fig. 12.5 ADH and ALDH polymorphisms that affect alcohol metabolism

Table 12.2 Polymorphisms of detoxification enzymes

Detoxification enzymes	Genotype
Transferases	
Glutathione S-transferase	GSTM1, GSTM1 mull
UDP-glucuronosyltransferase	UGT1A, UGT2A, UGT2B
Sulfotransferase	SULT1A1*1, SULT1A1*2
N-acetyltransferase	NAT1, NAT2
Methyltransferase	COMTVal158, COMTVal158Met
Non-transferases	
Quinone oxidoreductase	NQO1C609, NQO1C609T
Epoxide hydrolase	EPHX1T113C, EPHX1A139G

12.3 Genetic Polymorphisms of Detoxification Enzymes

Metabolic enzyme–catalyzed reactions serve as activation and detoxification processes to make foreign compounds and endogenous metabolites more soluble in water, thus facilitating their excretion from the body. Phase II detoxification enzymes refer to the enzymes that catalyze the detoxification reactions of foreign compounds, including glutathione S-transferase (GST), UDP-glucuronosyltransferase (UGT), N-acetyltransferase (NAT), and sulfotransferase (SULT).

Table 12.2 lists genetic polymorphisms of detoxification enzymes associated with foreign compound–mediated cytotoxicity and carcinogenesis. These enzymes are important in detoxifying xenobiotics to protect tissues from oxidative stress damages to cellular macromolecules, including proteins, lipids, and nucleic acids.

12.3 Genetic Polymorphisms of Detoxification Enzymes

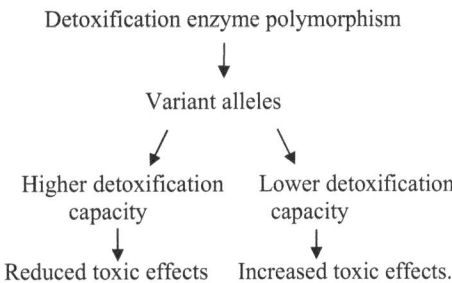

Fig. 12.6 Effects of detoxification enzyme polymorphisms on foreign compound–mediated toxic effects

Detoxification enzyme polymorphisms affect foreign compounds–mediated toxic effects. For example, differences in the levels of expression and catalytic activities of GST between individuals have been reported in the detoxification of polycyclic aromatic hydrocarbons.

Figure 12.6 presents the effects of detoxification enzyme polymorphisms on foreign compound–mediated toxic effects. An allele is one of the forms of gene that appears at a particular location on a specific chromosome. A variant allele with a higher detoxifying capacity can lead to reduced toxic effects, while that with a lower detoxifying capacity can result in increased toxic effects.

12.3.1 Glutathione S Transferase (GST)

Glutathione S-transferase (GST) genes encode a family of detoxification enzymes that are involved in the detoxification metabolism by conjugating foreign compounds with glutathione. GST catalyzes the addition of the thiol of GSH to electrophilic acceptors. By detoxifying a variety of electrophilic compounds, GST protects cells from reactive oxygen species–induced damages to intracellular molecules such as proteins, lipids, and nucleic acids.

Genetic polymorphisms in GTS enzymes can alter the catalytic efficiency of GST, resulting in variations in their detoxification activities. Consequently, genetic variations in detoxification enzymes influence the susceptibility of an organism to xenobiotic toxic effects. GSTT1 and GSTM1 are two relevant GST polymorphisms that play an important role in the detoxifying reactive oxygen species involved in the cellular processes of inflammation, aging, and degenerative diseases.

GSTM1 null polymorphisms result in an increased oxidative stress in comparison with wild genotype, which lead to significantly higher DNA damage. GSTM1 and GSTT1-deleted polymorphisms were found to exhibit higher risk of developing various diseases. The association between GSTM1 and GSTT1 genotypes and health risk has been observed in the lung. It was also reported that the null genotypes of GSTT1 and GSTM1 genes may be associated with an increased risk of cancer. GSTM1 homozygous null genotype may also lead to an increased risk of hypertension. In addition, studies also suggested that GSTT1 and GSTM1 deletion genotypes are associated with pathological conditions that develop age-related cataract.

12.3.2 UDP-Glucuronosyltransferase

UDP-glucuronosyltransferases (UGTs) are a major metabolizing enzyme involved in the detoxification of a large number of drugs, environmental chemicals, and endogenous compounds. Glucuronidation reactions catalyzed by UGT enzymes attributes to a detoxification process that adds glucuronic acid to exogenous and endogenous compounds, thus reducing the extent of their toxicity and facilitating their excretion from the body.

Glucuronidation is a crucial pathway of metabolism that detoxifies reactive chemical intermediates including reactive oxygen species. UGT function is genetically modulated by nucleotide polymorphisms, which lead to the expression of functionally altered enzymes. Due to genetic polymorphisms and differences in the regulation of UGT genes, the expression of UGT varies widely between individuals.

Genetic polymorphisms in UGT may result in variations in its detoxification activities and subsequently influence the levels of toxicants. UGTs are expressed predominantly in the liver and gastrointestinal tract. UGTs contribute to the detoxification of drugs by forming water-soluble glucopyranosiduronic acids. Consequently, modulation of UGT enzymatic activity or gene promoter activity has been implicated as a risk factor for adverse drug effects.

UGT-catalyzed glucuronidation is a primary detoxification pathway of dietary heterocyclic amines and major tobacco carcinogens, polycyclic aromatic hydrocarbons. Genetic polymorphisms in genes that encode enzymes involved in metabolizing tobacco carcinogens could affect an individual's risk for lung cancer.

UGT consists of two subfamilies, UGT1 and UGT2. One of the subfamilies of UGT1 enzyme is UGT1A. The functional UGT1A isoforms are encoded by a single UGT1A gene locus with multiple exons. UGT1A1 plays a significant role in the detoxification of estrogens. Catechol-estrogen metabolites can induce carcinogenesis by acting as endogenous tumor initiators. Glucuronidation, mediated by UGT1A1 enzyme, is a main metabolic pathway of estrogen detoxification in steroid target tissues.

Individual susceptibility to colorectal cancer may be in part due to variations in the detoxification capacity of subfamilies of UGT1. Variant UGT1A6 and UGT1A7 genotypes with reduced enzymatic activities might enhance susceptibility. UGT1A6 may also be used to identify people with increased risk of developing lung cancer. While UGT2 family is subdivided into two subfamilies, UGT2A and UGT2B. Their functional polymorphisms may play a role in tobacco-related cancer risk.

12.3.3 Sulfotrasferase

Sulfotransferase (SULT) catalyzes sulfate conjugation of various phenolic derivatives of foreign compounds including drugs and other chemicals as well as endogenous substances such as neurotransmitters and steroid hormones. Genetic polymorphisms in three alleles of human SULT1A1 gene are SULT1A1*1, SULT1A1*2, and SULT1A1*3.

12.3 Genetic Polymorphisms of Detoxification Enzymes

Marked differences in the activities of SULT1A1 variants toward a variety of substrates are critical for understanding interindividual variability of SULT in response to toxicants and carcinogens. SULT1A1 is involved in the bioactivation of N-hydroxy metabolite of dietary procarcinogen heterocyclic amines. The variant allele of the SULT1A1 gene may be associated with lung cancer. Moreover, comparative studies of SULT1A1 gene polymorphism reveal that variations in SULT1A1 activity may attribute to prostate cancer risk.

The low activity SULT1A1*2 allele may be at higher risk following carcinogen exposure than that with the SULT1A1*1 genotype. Individuals with the lower activity SULT1A1*2 allele may be at a higher cancer risk following carcinogen exposure than those with the higher activity SULT1A1*1 genotype. While SULT1A1*2 polymorphism enhances the effect of endogenous estrogen exposure, suggesting that this allele polymorphism may be a risk factor for breast cancer.

12.3.4 N-Acetyltransferase

N-Acetyltransferases (NATs)-catalyzed acetylation reactions play an essential role in the detoxification of aromatic and heterocyclic amines in foreign compounds, including therapeutic drugs and environmental chemicals. The acetylation polymorphism is a common inherited variation in human drug and carcinogen metabolism. Genetic polymorphisms in NATs may act as molecular biomarkers to provide important predictive information about cytotoxicity and carcinogenesis.

NAT1 and NAT2 are important in detoxifying carcinogenic amines. Amines undergo metabolism through N- or O-acetylation pathway. Studies of NAT polymorphisms revealed a significance association of slow acetylation phenotype with bronchial and asthma. Polymorphisms in NAT2 have been suggested as susceptibility factors for DNA damage and lung cancer. Lung adduct levels are significantly increased for the NAT2 slow acetylation genotype.

The combination of NAT2 slow genotype with glutathione S-transferases GSTM1 null genotype noticeably enhances the levels of lung adducts, suggesting a synergistic effect on DNA damage in the lung tissue. Many studies have focused on genetic variation in NAT2 and its potential as a risk factor in colorectal cancer. Moreover, women who smoke and have a genetically NAT2 slow acetylation genotype or GSTM1 null genotype are at increased risk of breast cancer.

12.3.5 Methyltransferase

Catechol-O-methyltransferase (COMT) catalyzes the transfer of a methyl group from S-adenosylmethionine to a catechol-containing substrate molecule. COMT methylation is a metabolic inactivation pathway for catechol estrogens. Intensive studies have revealed a role for the oxidative metabolites of estrogen, particularly catechol estrogens, in the development of estrogen carcinogenesis.

COMT is polymorphic in human populations. COMT gene polymorphisms have been shown to affect COMT enzyme functions. The variant form of COMT can affect the detoxification efficiency of endogenous or exogenous catechol estrogens. The COMT Val158Met polymorphism, where valine is substituted with methionine at nucleotide position 158, may change the secondary or tertiary structure of COMT, making it more susceptible to undergo inhibition and inactivation. The variant Val158Met in the COMT gene was found to exhibit a several-fold decrease in enzymatic activity.

A low activity of COMT variant can give rise to the accumulation of potentially carcinogenic catechol estrogens and their reactive intermediates, thus causing an increase in the risk of tumorigenesis. It has been reported that a low activity variant of the COMT enzyme is associated with an increased risk of developing breast cancer. Besides the gene, environment interaction is also crucial in modulating the physiological role of the COMT enzyme.

A reduction in metabolizing capacity of detoxification enzymes can lead to increased toxic effects from oxidative metabolites of estrogen. If not inactivated by COMT, catechol estrogens can generate a significant quantity of reactive oxygen species, causing oxidative stress and damages to health. In addition, SULT and UGT genes are also involved in catechol estrogen detoxification through conjugation reactions. Estrogen has been suggested to trigger breast cancer development via an initiating mechanism involving its metabolite. These findings supported a hypothesis that breast cancer can be initiated by estrogen exposure and that increased estrogen exposure confers a higher risk of breast cancer.

12.3.6 Quinone Oxidoreductase

Nontransferase detoxification enzymes include quinone oxidoreductase and epoxide hydrolase. Table 12.2 lists genetic polymorphisms of these nontransferase enzymes. NADH:quinone oxidoreductase 1 (NQO1) is a flavoprotein that catalyzes detoxification reduction by converting quinoid compounds to hydroquinones, thus protecting cells from cytotoxicity, carcinogenicity, and oxidative stress.

NQO1 enzyme also detoxifies environmental carcinogens such as nitroaromatic compounds and heterocyclic amines. Genetic polymorphisms of NQO1 may result in variations in its detoxification activities, leading to influencing the levels of toxicant and carcinogenic compounds present in the body.

The gene coding for NQO1 has a genetic polymorphism C609T in which cytosine is substituted with thymine at nucleotide position 609 of the NQO1 cDNA. NQO1 C609T polymorphism reduces NQO1 enzyme activity, thereby diminishing the detoxifying capacity provided by the enzyme. A genetic variant in NQO1 may modify individual susceptibility to hepatocellular carcinoma. The reduced enzymatic activity in NQO1 C609T polymorphism may play a pivotal role in tumor development. A significantly greater risk of hepatocellular carcinoma in association with the variant NQO1 C609T implies that NQO1 C609T genotype may affect individual susceptibility to liver cancer.

12.3.7 Epoxide Hydrolase

Microsomal epoxide hydrolase (EPHX) is a nontransferase detoxification enzyme that detoxifies chemicals such as butadiene, benzene, and styrene. EPHX also plays an important role in the activation and detoxification of polycyclic aromatic hydrocarbons present in tobacco smoking and grilled meat. The enzyme genetic polymorphisms may serve as molecular biomarkers to provide important information on carcinogenesis.

As a pathway to metabolize polycyclic aromatic hydrocarbons, EPHX polymorphism may be considered as a genetic susceptibility factor for lung cancer and for the development of esophageal carcinogenesis. A significant correlation was found between EPHX functional polymorphisms, T113C and A139G, and the individual's susceptibility to pulmonary disease.

12.4 Enzyme Polymorphisms on Alcohol and Smoke

The liver enzymes, alcohol dehydrogenase (ADH) and aldehyde dehydrogenase (ALDH), are two important enzymes involving alcohol metabolism. Both enzymes are responsible for the oxidative metabolism of ethanol. ADH catalyzes oxidation of ethanol to acetaldehyde, while ALDH oxidizes acetaldehyde to acetate, which is further metabolized to CO_2 and H_2O. Genetic polymorphisms of ADH and ALDH have an influence on alcoholism susceptibility. The kinetics of ethanol absorption and elimination is influenced by genetic factors.

Similarly, enzyme polymorphisms are also involved in addict to smoking. Blood cultures obtained from GSTT1 and GSTM1 null individuals were found to exhibit an increase in the sensitivity to smoke toxins. Individuals who carry GSTM1-null genotype were reported to have a higher susceptibility to DNA damage induced by tobacco smoke than GSTM1-positive ones.

12.4.1 Alcoholism

Genetic polymorphisms of the enzymes alcohol dehydrogenase (ADH) and aldehyde dehydrogenase (ALDH) are associated with alcohol drinking habits. Genetic variations in ADH and ALDH among individuals in response to alcohol have been reported to contribute to differences in susceptibility to alcohol abuse and dependence, particularly the functional genetic variants of ADH that exhibit low alcohol oxidizing activity. While the genetic variants of ALDH that exhibit low acetaldehyde oxidizing activity protect against heavy drinking and alcoholism.

ADH1B and ADH1C alleles encode active ADH enzymes, resulting in more rapid conversion of alcohol to acetaldehyde. These alleles have a protective effect on the risk of alcoholism. Moreover, a variant of the ALDH2 gene encodes an essentially inactive ALDH enzyme, resulting in acetaldehyde accumulation and a protective effect.

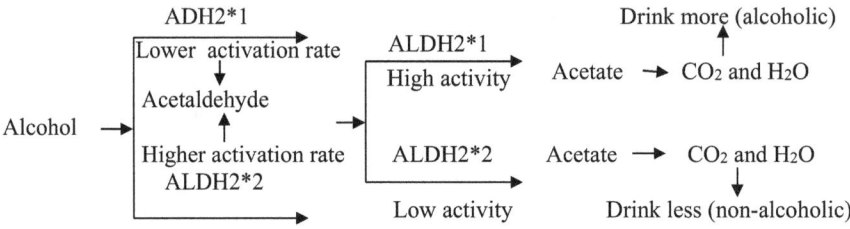

Fig. 12.7 ADH and ALDH polymorphisms affect alcohol metabolisms

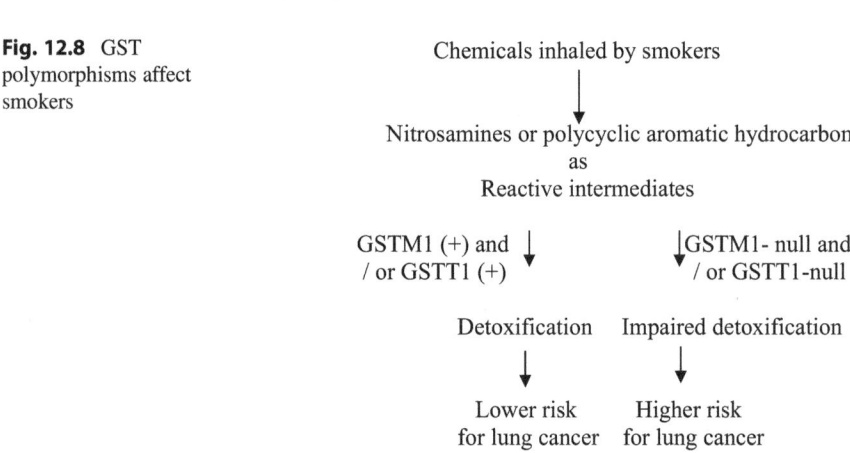

Fig. 12.8 GST polymorphisms affect smokers

Figure 12.7 shows that polymorphic variants in ADH and ALDH affect alcohol metabolism. The alcoholics were found to have significantly higher frequencies of ALDH2*1 allele than did the non-alcoholic subjects.

12.4.2 Smoker

A higher level of DNA adducts has been detected in lymphocytes of tobacco smokers who lack the GSTM1 gene. Moreover, several studies have also revealed an association of the homozygous null deletions in GSTM1 and GSTT1 with an increasing risk of lung cancer. Figure 12.8 as shown below reveals that polymorphic variants in GST affect the risk of smokers to cancer.

12.5 Species Difference in Metabolic Enzyme Activities

Variations in metabolic enzyme activity are often the basis for species differences in an organism's susceptibility to foreign compound–mediated toxicity. A number of studies have been reported to support the proposal that metabolic differences are the underlying cause of species variation in carcinogenicity. Such differences in carcinogenicity may also include variations in metabolic pathways between species.

12.5 Species Difference in Metabolic Enzyme Activities

It has been reported that the susceptibility to aflatoxin, tamoxifen, and 4-ipomeanol toxicities are different between humans and mice. These observed differences are mainly attributed to species variations in the activities of metabolic enzymes that catalyze detoxification reactions for these compounds.

12.5.1 Susceptibility to Aflatoxin Toxicity between Humans and Mice

Aflatoxin B_1 is a major risk factor for hepatocellular carcinoma in humans. Activation of aflatoxin B_1 catalyzed by CYP3A4 forms a carcinogenic intermediate, aflatoxin-8,9-epoxide, which is detoxified by the conjugation reaction catalyzed by glutathione S-transferase A3 (GSTA3). GSTA3-catalyzed conjugation reaction forms glutathione-conjugated aflatoxin derivative to facilitate the excretion of aflatoxin B_1.

GSTA3 subunit plays a crucial role in protection against aflatoxin B_1 toxicity. Due to relatively low expression of GSTA3 activity for aflatoxin-8,9-epoxide, humans are highly susceptible to toxic and carcinogenic effects of aflatoxin B_1. Unconjugated aflatoxin-8,9-epoxide can react with DNA to form a DNA adduct.

Conversely, mice are resistant to aflatoxin B_1 carcinogenesis due to a high expression of mGSTA3. Such species difference in detoxifying aflatoxin B_1 between humans and mice is described in Fig. 12.9.

12.5.2 Resistance to Tamoxifen Toxicity in Humans, Not in Rats

Tamoxifen is an important antiestrogenic drug used for the treatment of breast cancer. Activation of tamoxifen catalyzed by CYP3A4 involves hydroxylation reaction, which produces oxidative metabolites, including hydroxytamoxifen derivatives.

Detoxification pathways for α-hydroxytamoxifen proceed in two distinctive catalytic reactions. One pathway is a conjugation reaction catalyzed by UDP-glucuronosyl-transferase (UGT), resulting in the formation of inactive glucuronidate

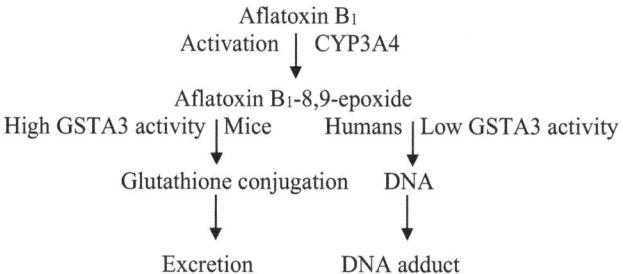

Fig. 12.9 Aflatoxin B_1 toxicity in humans, but not in mice

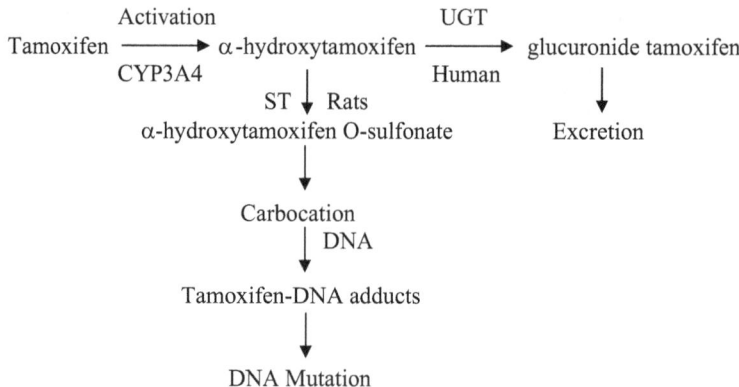

Fig. 12.10 Susceptibility to tamoxifen toxicity in rats and humans

tamoxifen. Another detoxification pathway is a conjugation reaction catalyzed by sulfotransferase (ST), which produces tamoxifen o-sulfonate.

Species differences in tamoxifen detoxification metabolism between humans and rats are presented in Fig. 12.10. Humans exhibit a much higher activity of UGT than rats, and have more effective glucuronylation of α-hydroxytamoxifen in liver microsomes than rats. Such high UGT activity plays a key role for humans to defend against the toxicity of tamoxifen.

In contrast, tamoxifen is a potent hepatocarcinogen in rats. Rats display a low activity of UGT and a high activity of ST. α-hydroxytamoxifen in rats prefer a conjugation reaction through sulfonation to glucuronidation, which leads to the formation of tamoxifen O-sulfonate, and subsequently releases its sulfonate group to produce reactive carbocation. Carbocation can react with DNA to form DNA adducts. Consequently, tamoxifen is a liver carcinogen in rats.

12.5.3 Different 4-Ipomeanol Toxicity Between Humans and Rodents

4-Ipomeanol is a naturally occurring furan, which was reported to induce pulmonary toxicity in the lungs of several mammalian species. To induce cytotoxicity to rodents, 4-Ipomeanol requires metabolic activation catalyzed by CYP4B1 to form a highly reactive furan epoxide.

While in humans, the risk is minimal for developing pulmonary toxicity. Instead of metabolic activation by CYP4B1 in pulmonary cells, 4-Ipomeanol is metabolically activated in the liver by hepatic enzymes, CYP3A4 and CYP1A2. The produced reactive furan epoxide causes selective hepatotoxicity in humans.

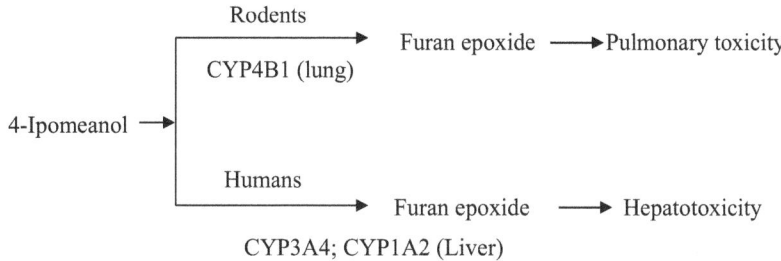

Fig. 12.11 Different 4-Ipomeanol toxicity between humans and rodents

Figure 12.11 illustrates distinctive responses to 4-Ipomeanol toxicity between humans and rodents. The figure shows that deviations in response to 4-Ipomeanol toxicity between humans and rodents are attributed to not only different sites of metabolism but also distinctive CYP450 isozymes present in the metabolic activation.

Bibliography

Abdel-Rahman SZ, Salama SA, Au WW et al (2000) Role of polymorphic CYP2E1 and CYP2D6 genes in NNK-induced chromosome aberrations in cultured human lymphocytes. Pharmacogenetics 10(3):239–249

Alvarez-Diez TM, Zheng J (2004) Mechanism-based inactivation of cytochrome P450 3A4 by 4-ipomeanol. Chem Res Toxicol 17:150–157

Angstadt AY, Berg A, Zhu J et al (2013) The effect of copy number variation in the phase II detoxification genes UGT2B17 and UGT2B28 on colorectal cancer risk. Cancer 119(13): 2477–2485

Archer MC (1981) Reactive intermediates from nitrosamines. Adv Exp Med Biol 136:1027–1035

Autrup H (2000) Genetic polymorphisms in human xenobiotic metabolizing enzymes as susceptibility factors in toxic response. Mutat Res 464(1):65–76

Badal S, Delgoda R (2014) Role of the modulation of CYP1A1 expression and activity in chemoprevention. J Appl Toxicol 34(7):743–753

Baer BR, Rettie AE, Henne KR (2005) Bioactivation of 4-ipomeanol by CYP4B1: adduct characterization and evidence for an enedial intermediate. Chem Res Toxicol 18:855–864

Bajro MH, Josifovski T, Panovski M et al (2012) Promoter length polymorphism in UGT1A1 and the risk of sporadic colorectal cancer. Cancer Genet 205(4):163–167

Baum M, Amin S, Guengerich FP et al (2001) Metabolic activation of benzo[c]phenanthrene by cytochrome P450 enzymes in human liver and lung. Chem Res Toxicol 14:683–693

Boccia S, Persiani R, La Torre G et al (2005) Sulfotransferase 1A1 polymorphism and gastric cancer risk: a pilot case-control study. Cancer Lett 229(2):235–243

Chen C-H (2012) Activation and detoxification enzymes: functions and implications. Springer Sciences, New York

Chen C-H (2020) Xenobiotic metabolic enzymes: bioactivation and antioxidant defense. Springer Nature, Switzerland

Clapper ML (2000) Genetic polymorphism and cancer risk. Curr Oncol Rep 2(3):251–256

Czerwinski M, McLemore TL, Philpot RM et al (1991) Metabolic activation of 4-ipomeanol by complementary DNA-expressed human cytochromes P-450: evidence for species-specific metabolism. Cancer Res 51:4636–4638

Daniels J, Kadlubar S (2013) Sulfotransferase genetic variation: from cancer risk to treatment response. Drug Metab Rev 45(4):415–422

Edenberg HJ, McClintick JN (2018) Alcohol dehydrogenases, aldehyde dehydrogenases, and alcohol use disorders: a critical review. Alcohol Clin Exp Res 42(12):2281–2297

Hayes JD, Strange RC (2000) Glutathione S-transferase polymorphisms and their biological consequences. Pharmacology 61(3):154–166

Hein DW (2002) Molecular genetics and function of NAT1 and NAT2: role in aromatic amine metabolism and carcinogenesis. Mutat Res 506-507:65–77

Hisamuddin IM, Yang VW (2007) Genetic polymorphisms of human flavin-containing monooxygenase 3: implications for drug metabolism and clinical perspectives. Pharmacogenomics 8(6):635–643

Kato S, Shields PG, Caporaso NE et al (1994) Analysis of cytochrome P450 2E1 genetic polymorphisms in relation to human lung cancer. Cancer Epidemiol Biomarkers Prev 3:515–518

Kim SY, Laxmi YR, Suzuki N et al (2005) Formation of tamoxifen-DNA adducts via O-sulfonation, not O-acetylation, of alpha-hydroxytamoxifen in rat and human livers. Drug Metab Dispos 33:1673–1678

Kiss I, Sándor J, Pajkos G, Bogner B et al (2000) Colorectal cancer risk in relation to genetic polymorphism of cytochrome P450 1A1, 2E1, and glutathione-S- transferase M1 enzymes. Anticancer Res 20:519–522

Landerer S, Kalthoff S, Paulusch S, Strassburg CP (2020) UDP-glucuronosyltransferase polymorphisms affect diethylnitrosamine-induced carcinogenesis in humanized transgenic mice. Cancer Sci 111(11):4266–4275

Leung TM, Lu Y (2017) Alcoholic liver disease: from CYP2E1 to CYP2A5. Curr Mol Pharmacol 10(3):172–178

Li TK (2000) Pharmacogenetics of responses to alcohol and genes that influence alcohol drinking. J Stud Alcohol 61:5–12

Li H, Fu WP, Hong ZH (2013) Microsomal epoxide hydrolase gene polymorphisms and risk of chronic obstructive pulmonary disease: a comprehensive meta-analysis. Oncol Lett 5(3):1022–1030

Mackenzie PI, Gregory PA, Lewinsky RH et al (2005) Polymorphic variations in the expression of the chemical detoxifying UDP glucuronosyltransferases. Toxicol Appl Pharmacol 207(2 Suppl):77–83

Merali Z, Ross S, Paré G (2014) The pharmacogenetics of carboxylesterases: CES1 and CES2 genetic variants and their clinical effect. Drug Metabol Drug Interact 29(3):143–151

Miller DP, De Vivo I, Neuberg D et al (2003) Association between self-reported environmental tobacco smoke exposure and lung cancer: modification by GSTP1 polymorphism. Int J Cancer 104(6):758–763

Neafsey P, Ginsberg G, Hattis D et al (2009) Genetic polymorphism in CYP2E1: population distribution of CYP2E1 activity. J Toxicol Environ Health B Crit Rev 12:362–388

Nimura Y, Yokoyama S, Fujimori M et al (1997) Genotyping of the CYP1A1 and GSTM1 genes in esophageal carcinoma patients with special reference to smoking. Cancer 80:852–857

Oscarson M (2001) Genetic polymorphisms in the cytochrome P450 2A6 (CYP2A6) gene: implications for interindividual differences in nicotine metabolism. Drug Metab Dispos 29:91–95

Palma S, Cornetts T, Padua L et al (2007) Influence of glutathione S- transferase polymorphisms on genotoxic effects induced by tobacco smoke. Mutat Res 633:1–12

Phillips IR, Shephard EA (2019) Flavin-containing monooxygenase 3 (FMO3): genetic variants and their consequences for drug metabolism and disease. Xenobiotica 18:1–60

Phillips DH, Carmichael PL, Hewer A et al (1996) Activation of tamoxifen and its metabolite alpha-hydroxytamoxifen to DNA-binding products: comparisons between human, rat and mouse hepatocytes. Carcinogenesis 17:89–94

Sak K (2017) The Val158Met polymorphism in COMT gene and cancer risk: role of endogenous and exogenous catechols. Drug Metab Rev 49(1):56–83

Saldivar SJ, Wang Y, Zhao H et al (2005) An association between a NQO1 genetic polymorphism and risk of lung cancer. Mutat Res 582(1–2):71–78

Shupe T, Sell S (2004) Low hepatic glutathione S-transferase and increased hepatic DNA adduction contribute to increased tumorigenicity of aflatoxin B1 in newborn and partially hepatectomized mice. Toxicol Lett 148:1–9

Sireesha R, Laxmi SGB, Mamata M et al (2012) Total activity of glutathione-S-transferase (GST) and polymorphisms of GSTM1 and GSTT1 genes conferring risk for the development of age-related cataracts. Exp Eye Res 98:67–74

van der Logt EM, Bergevoet SM, Roelofs HM et al (2004) Genetic polymorphisms in UDP-glucuronosyltransferases and glutathione S-transferases and colorectal cancer risk. Carcinogenesis 25(12):2407–2415

Zanger UM, Schwab M (2013) Cytochrome P450 enzymes in drug metabolism: regulation of gene expression, enzyme activities, and impact of genetic variation. Pharmacol Therapeu 138:103–141

Vogel A, Ockenga J, Ehmer U et al (2002) Polymorphisms of the carcinogen detoxifying UDP-UGT1A7 in proximal digestive tract cancer. Z Gastroenterol 40(7):497–502

Waring RH (2020) Cytochrome P450: genotype to phenotype. Xenobiotica 50(1):9–18

Werk AN, Cascorbi I (2014) Functional gene variants of CYP3A4. Clin Pharmacol Ther 96(3):340–348

Yamazaki H (2017) Differences in toxicological and pharmacological responses mediated by polymorphic cytochromes P450 and related drug-metabolizing enzymes. Chem Res Toxicol 30(1):53–60

Yoda T, Tochitani T, Usui T et al (2022) Involvement of the CYP1A1 inhibition-mediated activation of aryl hydrocarbon receptor in drug-induced hepatotoxicity. J Toxicol Sci 47(9):359–373

Defense Against Oxidative Stress: Nrf2-ARE Pathway

Reactive chemical intermediates, including reactive oxygen and nitrogen species and free radicals, are formed during foreign compound metabolic conversions catalyzed by metabolic enzymes. In many cases, such reactive chemical intermediates lead to oxidative stress, as the production of reactive chemical intermediates exceeds the ability of metabolic detoxification enzymes and endogenous antioxidants to neutralize them.

Such an imbalance can cause cell damages. Reactive chemical intermediates can attack cellular components (proteins, membrane lipids, and nucleic acids), leading to cytotoxicity, lipid peroxidation, and carcinogenesis. To maintain a balance between bioactivation and detoxification, living cells maintain an inducible antioxidant response pathway.

Figure 13.1 shows an imbalance between the production of metabolic reactive intermediates and the body's defense mechanism, resulting in oxidative stress. Inflammation is a biological response of oxidative stress, while oxidative stress is a biochemical dysregulation. Chronic inflammation is strongly associated with the major stages of cytotoxicity and carcinogenesis.

13.1 Transcription Factor Nrf2

The activity of antioxidant response pathway is largely regulated by nuclear factor erythroid-2-related factor 2 (Nrf2). Nrf2 activation mechanism occurs as Nrf2 dissociates from its inactive complex with the repressor protein Keap1 and subsequently translocate into the nucleus. Once in the nucleus, Nrf2 complex binds to antioxidant response elements (ARE), promoting transcription of cytoprotective genes. Nrf2-ARE-dependent genes include NAD(P)H:quinone oxidoreductase, glutathione-S-transferase, UDP glucuronosyltransferase, and other detoxification enzymes.

Fig. 13.1 Oxidative stress and cytotoxicity

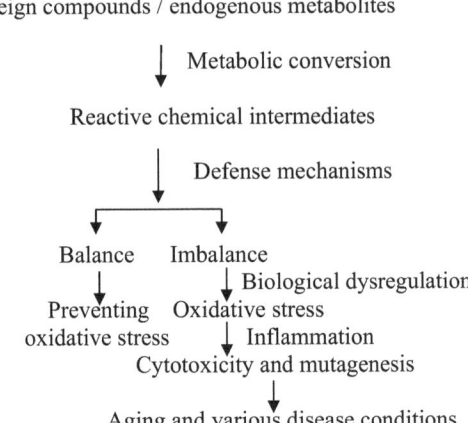

Nrf2 is an important chemoprevention and therapeutic target. It governs the expression of many detoxification enzymes and antioxidants, acts as a sensor of oxidative stress, and prevents genome instability. The body utilizes such antioxidative stress defense systems in the cells to neutralize or eliminate chemical reactive species. The induction of phase II detoxification enzymes is an important defense mechanism against reactive chemical intermediates. Nrf2 regulates the expression of antioxidants to protect the body against oxidative stress and its induced inflammation.

13.1.1 Role of Nrf2 on Oxidative Stress

Oxidative stress is a result of an imbalance between the production of metabolic reactive intermediates and the body's defense mechanism. Its associated cellular damages include protein adducts, lipid peroxidation, and DNA mutagenicity. Such cellular damages are fundamental pathological processes in a variety of disease conditions. Increasing evidences also reveal that oxidative stress and inflammation are common pathological mechanisms underlying neurodegeneration.

Nrf2 governs the expression of many detoxification enzymes and antioxidants, and acts as a sensor of oxidative stress. The body utilizes antioxidative stress defense systems in the cells to neutralize or eliminate reactive intermediate species. Hence, to maintain a balance between bioactivation and detoxification, it is important that living cells maintain the important antioxidant response pathway regulated by Nrf2. Nrf2 is known for its ability to regulate the expression of a series of metabolic enzymes, such as glutathione S-transferase, UDP-glucuronosyltransferase, and NAD(P)H quinone oxidoreductase.

13.1.2 Keap1 Regulation of Nrf2 Activity

In response to oxidative stress, an intricate molecular mechanism facilitated by sensor cysteines within Keap1 allows Nrf2 to escape ubiquitination, accumulate within the cell, and translocate to the nucleus, where it can promote its antioxidant transcription program. Under homeostatic conditions, Keap1 forms part of an E3 ubiquitin ligase, which tightly regulates the activity of the transcription factor Nrf2 by targeting it for ubiquitination and proteasome-dependent degradation.

The Keap1-Nrf2 pathway is the principal protective response to oxidative and electrophilic stresses. During oxidative stress, the Nrf2 - Keap1 interaction is disrupted, which allows Nrf2 to translocate to the nucleus, where Nrf2 complex binds to antioxidant response elements (ARE), promoting transcription of cytoprotective genes.

13.2 Activation of Nrf2-ARE Pathway

As pointed out above, the Nrf2-ARE pathway is an intrinsic mechanism of defense against oxidative stress. It regulates the expression of many genes and functions related to antioxidant proteins and detoxification enzymes. Such regulation processes function to eliminate toxicants and carcinogens, before they can cause damages to the cells. Activation of the Nrf2-ARE pathway is therefore critical to the elimination of reactive oxygen species, free radicals, and other reactive chemical intermediates.

Evidences point to oxidative stress and inflammation as a pathological mechanism underlying disease condition. Targeting the Nrf2-ARE pathway may also present a novel therapeutic approach for the treatments of inflammatory diseases as well as aging-related neurodegeneration and other disease conditions. Many studies showed a critical role of Nrf2 in cellular protection and anti-carcinogenicity, implying that the Nrf2-ARE pathway may serve as a therapeutic target for disease conditions.

For example, activating antioxidant systems in cells through the Nrf2-ARE pathway has been reported to limiting cardiac injury and heart failure. On the other hand, over activation of the Nrf2-ARE pathway was reported to cause cardiovascular injury. Moreover, over activation of Nrf2 in cancer cells has been implicated in cancer progression and in resistance to cancer chemotherapeutics.

13.3 Nrf2-KeapP1-ARE Pathway

Nrf2 activation is based on the dissociation of Nrf2 from its inactive complex with the repressor protein Keap1 and the subsequent translocation of Nrf2 into the nucleus. One possible mechanism in the defense against oxidative stress is the induction of Nrf2 by small molecule activators to modify Keap1 cysteine residues and to disrupt the Nrf2-Keap1 complex, thus favoring the dissociation of Nrf2 and subsequent nuclear translocation.

Another possible mechanism is the involvement of inhibiting Nrf2 ubiquitination by Keap1 and degrading by the proteasome to enhance the availability of Nrf2, thus favoring nuclear translocation of Nrf2.

Accordingly, the primary mechanism of Nrf2 activation is essentially based on the following three steps: (a) the dissociation of Nrf2 from its inactive complex with the repressor protein Keap1, (b) the subsequent translocation of Nrf2 into the nucleus, and (c) the activation of Nrf2 signaling, which induces ARE-dependent expression of detoxification enzymes and antioxidant proteins.

13.4 Molecular Mechanism of Nrf2-ARE Pathway

As described above, the Nrf2-ARE signaling pathway regulates the expression of genes which protein products are involved in the detoxification and elimination of reactive intermediates during bioactivation of foreign compounds. The Nrf2-Keap1-ARE pathway in the absence and presence of oxidative stress is described below:

13.4.1 In the Absence of Oxidative Stress

Figure 13.2 illustrates the Nrf2-Keap1-ARE pathway in the absence of oxidative stress. Under the nonstressed condition, Nrf2 covalently binds to cysteine residue of repressor Keap1 to form the complex Nrf2-Keap1 in the cytoplasm. Such Nrf2-Keap1 binding step is followed by the addition of ubiquitin (Ub) molecules to Nrf2, where Keap1 acts as an essential active protein.

In Fig. 13.2, the substrate for ubiquitination is highly specific. Its activity is provided by ubiquitin ligases, where Cullin 3 maintains a staging for ubiquitin ligases. The addition of ubiquitin molecules is referred to as ubiquitination, which affects Nrf2 transcription factor by initialing it for degrading unneeded Nrf2 via

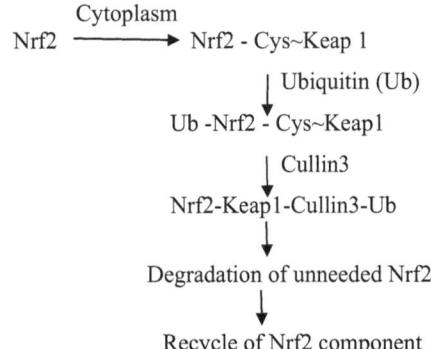

Fig. 13.2 Nrf2-Keap1-ARE pathway in the absence of oxidative stress

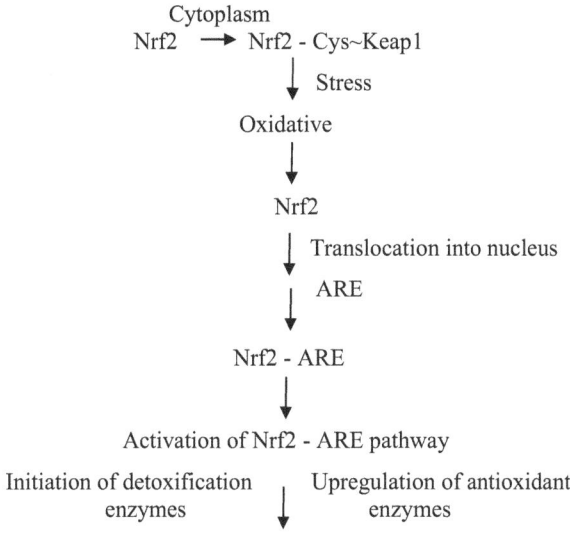

Fig. 13.3 Nrf2-Keap1-ARE Pathway in the presence of oxidative stress

proteasomes. In the absence of oxidative stress, Nrf2 is constantly degraded through ubiquitination and the degraded components are recycled in nonstressed cells.

13.4.2 In the Presence of Oxidative Stress

In the presence of oxidative stress, cysteine residues on Keap1 are modified, resulting in the stabilization and the translocation of Nrf2 into the nucleus. In the nucleus, Nrf2 binds to the promoter region of ARE, which initiates the transcription of cytoprotective detoxification enzymes and upregulates antioxidant function. Figure 13.3 illustrates the Nrf2-Keap1-ARE pathway in the presence of oxidative stress.

Following the exposure of Keap1 to oxidative stress, the Keap1-Cullin 3-ubiquitination system is disrupted and Nrf2 is not degraded. Upon the nuclear entry, Nrf2 binds to ARE in the promoter region of antioxidative genes, which initiates the transcription of antioxidative genes and the corresponding proteins.

13.5 Cytoprotection Through Nrf2-ARE Pathway

To prevent cellular damages caused by cytotoxicity and carcinogenic injury, it is important to reduce reactive chemical intermediates, including reactive oxygen and nitrogen species and free radicals. Cytoprotection is considered as a good strategy to reduce oxidative stress. Induction of Nrf2-ARE mediated expressions of antioxidants and detoxification enzymes with cytoprotective agents may counteract the toxic effects of reactive chemical intermediates.

In the liver, Nrf2 regulates the cytoprotective genes, which confer cellular resistance to toxicity and oxidative stress, therefore providing protection against inflammation and various disease conditions.

13.5.1 Induction of Nrf2 – ARE Pathway

Metabolic activation and detoxification play a critical role in the bioactivation, detoxication, and excretion of foreign compounds from the body. In many cases, activation reactions during biotransformation result in the generation of reactive intermediates, including reactive oxygen species and free radicals, and many of them may result in cytotoxicity and carcinogenicity. To counteract such harmful reactive chemical intermediate effects, living cells evolve sophisticated signaling mechanisms in response to oxidative stress.

Detoxification enzymes have long been investigated for potentially preventative and protective effects against oxidative stress-related cytotoxicity and carcinogenicity. Induction and activation of antioxidants and detoxification enzymes through the Nrf2-ARE signaling pathway represents a critical aspect of the cytoprotective mechanism of action.

Research has revealed that dietary phytochemicals possess the potential to protect organisms against cellular damages caused by oxidative stress because of their capacities to induce antioxidants and detoxification enzymes. Figure 13.4 illustrates of the Nrf2-ARE pathways as a strategy for cytoprotection of cells against damages caused by oxidative stress.

13.5.2 Enzyme Inducers for Cytoprotection

One of the major mechanisms of chemical protection against carcinogenesis, mutagenesis, and other forms of toxicity mediated by electrophiles is the induction of enzymes involved in their metabolism, particularly phase II enzymes, such as glutathione S-transferases, uridine diphosphate-glucuronosyltransferases, and

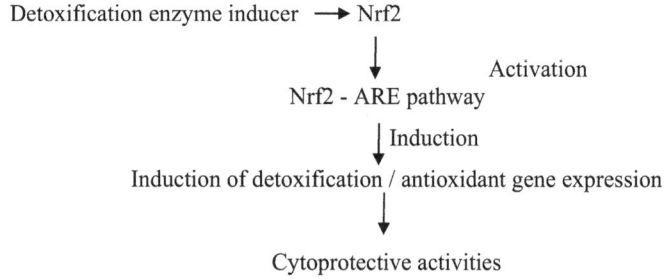

Fig. 13.4 Induction of Nrf2-ARE pathway for cytoprotection

13.5 Cytoprotection Through Nrf2-ARE Pathway

NAD(P)H:quinone reductase. Nrf2-ARE-dependent genes include these detoxification enzymes.

A strategy for chemoprevention is focused on the use of inducers to modulate the metabolism of foreign compounds, including drugs and other chemicals, through upregulation of metabolic detoxification enzymes. Figure 13.4 illustrates the induction of the Nrf2-ARE pathway for cytoprotection.

Detoxification enzyme inducers, such as phytochemicals, are compounds mainly produced by plants and widely distributed in fruits and vegetables. For example, phytochemical quercetin increases NAD(P)H quinone dehydrogenase 1 transcription, suggesting that dietary plant polyphenols can stimulate the transcription of detoxification enzymes, thus offering a protection against carcinogenic chemicals.

Moreover, dietary curcumin was found to induce the activities of glutathione S-transferase and NADH:quinone oxidoreductase, leading to increased detoxification of toxic benzo[a]pyrene. Quercetin and phenylpropanoids in ginger are also known chemoprotective compounds. They can significantly increase the level of Nrf2 activity. Anthocyanins among phytochemical flavonoids were reported to stimulate the antioxidant defense system, leading to affecting the expression of detoxification enzymes and the modulation of antioxidative stress defense system.

In addition, dibenzoylmethane, a minor constituent of licorice, is a beta-diketone phytochemical with a wide variety of anti-cancer effects. Dibenzoylmethane was found to increase the DNA binding activity of Nrf2 and mediate the induction of detoxification enzymes so as to inhibit the formation of benzo[a]pyrene-induced DNA adducts.

Furthermore, licorice from the root of plant glycyrrhiza uralensis has been used to soothe gastrointestinal problems and to repair stomach lining through its anti-inflammatory properties. Other dietary phytochemicals include zerumbone, a sesquiterpene compound occurring in tropical ginger that has been implicated as a promising chemoprotective agents against colon and skin cancers.

13.5.3 Over-Activation of Nrf2-ARE Pathway

The Keap1-Nrf2-ARE pathway helps prevent inflammation and disease conditions caused by oxidative stress. Activation of Nrf2 signaling induces ARE-dependent expressions of detoxification and antioxidant defense proteins, which are implicated in helping various disease conditions. However, over-activation of Nrf2 in cells has been implicated in disease progression as well as in resistance to chemotherapeutics.

For example, over-activation of the Nrf2 pathway can induce harmful effects on neurodegenerative conditions. It was reported that specific inhibition of Keap1 can prevent neuronal toxicity in correlation with Nrf2 activation. An understanding of the mechanisms mediating Nrf2 inhibition may benefit research for preventing neurodegenerative conditions with minimal side-effects.

13.6 Role of Nrf2 in Diseases

Oxidative stress as a result of elevated levels of reactive oxygen species has been observed in various cancers. Elevated expression of Nrf2 target genes confers advantages in terms of stress resistance and cell proliferation in cells. Discovery and development of selective Nrf2 inhibitors could make a critical contribution to improving cancer conditions.

Studies reported that oxidative stress is also an essential component in the progression of diabetic kidney disease. The transcription factor Nrf2 plays critical roles in protecting the body against oxidative stress. Nrf2 contributes to the protection of the kidneys by suppressing oxidative stress and inflammation.

As discussed above, the levels of detoxification enzymes are regulated by a nuclear transcriptional factor Nrf2, which is sequestered in cytoplasm by Keap1. The transcription of ARE-driven genes is regulated in part by Nrf2. Exposure of cells to ARE inducers results in the dissociation of Nrf2 from Keap1 and facilitates translocation of Nrf2 to the nucleus. Hence, activation of Nrf2 enhances the levels of antioxidant and detoxification enzymes.

Reducing oxidative stress and chronic inflammation optimally requires an elevation of the levels of these enzymes. Detoxification enzymes remove potential carcinogens by converting them to harmless compounds for elimination from the body. Keap1-Nrf2 system plays an essential role in cellular protection by regulating many antioxidant and detoxification enzyme genes through the antioxidant response element.

Reactive oxygen intermediates and toxic electrophiles are the major causes of neoplastic and chronic degenerations. Induction of detoxification genes against the damage of electrophiles and reactive oxygen intermediates is potentially a major strategy for reducing the risk of such disease conditions.

For example, the main risk factor of neurodegenerative diseases is aging. The protective role of the Nrf2-ARE pathway in neurodegenerative conditions on the reduction of oxidative stress and neuroinflammation has drawn much attention. Parkinson's disease (PD) is a progressive neurodegenerative disorder. Oxidative stress plays a critical role in dopaminergic neuronal cell death in PD. Nrf2 activation through interaction with the Nrf2-Keap1 may be effective in improving PD conditions. Hence, effects of Keap1-Nrf2 is considered as a potential therapeutic target for Alzheimer's disease.

13.7 Nrf2-Inducing Compounds for Chemoprevention

Understanding the Keap1-Nrf2 system and the potential effects of phytochemicals are important in preventing chronic diseases and maintenance of human health. The induction of detoxification enzymes is an effective means for achieving protection against carcinogens. One major mechanism of such protection is the induction of enzymes involved in their metabolism, particularly phase II detoxification enzymes, such as glutathione S-transferases, UDP-glucuronosyl transferases, and quinone reductases.

Dietary phytochemicals widely distributing in fruits and vegetables have been considered to possess such chemoprevention potential. Many such inducers activate detoxification enzyme genes involved in detoxifying toxicants and carcinogens, leading to the attenuation of oxidative stress. Diets-derived compounds studied for cytoprotective activities include resveratrol from grapes and broccoli, curcumins from turmeric, flavonoid genistein from soy, and quercetin from onion and green tea. Other studied phytochemicals include polyunsaturated fatty acids, carotenoids, lycopene and lutein, indole-3-carbinol, isothiocyanates, and pomegranate.

For example, sulforaphane, an isothiocyanate derived from cruciferous vegetables, is the most extensively studied phytochemical in the literature. Sulforaphane reacts with protein thiols to form adducts, thus affecting cysteine residues in Keap1 protein and modulating Nrf2 protein stability. Isothiocyanate from horseradish was also reported to activate ARE, mediate Nrf2 binding to ARE, and induce detoxification enzyme genes.

The induction of detoxification enzyme genes was also mediated by major garlic components, such as diallyl sulfide, disulfide, and trisulfide. Among them, 1,2-dithiole-3-thione, the sulfur-containing dithiolethiones, is present in cruciferous vegetables. In addition, oltipraz, sulforaphane, isothiocyanate, and curcumin are also potential chemoprevention agents. Research was carried out on chemoprevention of aflatoxin B1-induced hepatocarcinogenesis by oltipraz.

Bibliography

Baird L, Yamamoto M (2020) The molecular mechanisms regulating the KEAP1-NRF2 pathway. Mol Cell Biol 40(13):e00099–e00020

Bayele HK, Debnam ES, Srai KS (2016) Nrf2 transcriptional derepression from Keap1 by dietary polyphenols. Biochem Biophys Res Commun 469:521–528

Bryan HK, Olayanju A, Goldring CE et al (2013) The Nrf2 cell defence pathway: Keap1-dependent and -independent mechanisms of regulation. Biochem Pharmacol 85(6):705–717

Buendia I, Michalska P, Navarro E (2016) Nrf2-ARE pathway: an emerging target against oxidative stress and neuroinflammation in neurodegenerative diseases. Pharmacol Ther 157:84–104

Chakkittukandiyil A, Sajini DV, Karuppaiah A et al (2022) The principal molecular mechanisms behind the activation of Keap1/Nrf2/ARE pathway leading to neuroprotective action in Parkinson's disease. Neurochem Int 156:105325

Chen C-H (2012) Activation and detoxification enzymes: functions and implications. Springer Sciences, New York

Chen C-H (2020) Xenobiotic metabolic enzymes: bioactivation and antioxidant defense. Chapter 13. Springer Nature, Switzerland

Johnson JA, Johnson DA, Kraft AD et al (2008) The Nrf2-ARE pathway: an indicator and modulator of oxidative stress in neurodegeneration. Ann N Y Acad Sci 1147:61–69

Joshi G, Johnson JA (2012) The Nrf2-ARE pathway: a valuable therapeutic target for the treatment of neurodegenerative diseases. Recent Pat CNS Drug Discov 7(3):218–229

Kaspar JW, Niture SK, Jaiswal AK (2009) Nrf2:INrf2(Keap1) signaling in oxidative stress. Free Radic Biol Med 47(9):1304–1309

Kensler TW, Curphey TJ, Maxiutenko Y et al (2000) Chemoprotection by organosulfur inducers of phase 2 enzymes: dithiolethiones and dithiins. Drug Metabol Drug Interact 17:3–22

Kerr F, Sofola-Adesakin O, Dobril K, Ivanov DK et al (2017) Direct Keap1-Nrf2 disruption as a potential therapeutic target for Alzheimer's disease. PLoS Genet 13(3):e1006593

Keum YS (2011) Regulation of the Keap1/Nrf2 system by chemopreventive sulforaphane: implications of posttranslational modifications. Ann N Y Acad Sci 1229:184–189

Kim S, Viswanath ANI, Jong-Hyun Park J-H et al (2020) Nrf2 activator via interference of Nrf2-Keap1 interaction has antioxidant and anti-inflammatory properties in Parkinson's disease animal model. Neuropharmacology 167:107989

Krajka-Kuźniak V, Paluszczak J, Baer-Dubowska W (2017) The Nrf2-ARE signaling pathway: an update on its regulation and possible role in cancer prevention and treatment. Pharmacol Rep 69(3):393–402

Kwak MK, Egner PA, Dolan PM et al (2001) Role of phase 2 enzyme induction in chemoprotection by dithiolethiones. Mutat Res 480-481:305–315

Lee JM, Johnson JA (2004) An important role of Nrf2-ARE pathway in the cellular defense mechanism. J Biochem Mol Biol 37:139–143

Lee JS, Surh YJ (2005) Nrf2 as a novel molecular target for chemoprevention. Cancer Lett 224(2): 171–184

Li W, Guo Y, Zhang C et al (2016) Dietary phytochemicals and cancer chemoprevention: a perspective on oxidative stress, inflammation, and epigenetics. Chem Res Toxicol 29:2071–2095

Li J, Calkins MJ, Johnson DA et al (2007) Role of Nrf2-dependent ARE-driven antioxidant pathway in neuroprotection. Methods Mol Biol 399:67–78

McWalter GK, Higgins LG, McLellan LI et al (2004) Transcription factor Nrf2 is essential for induction of NAD(P)H:quinone oxidoreductase 1, glutathione S-transferases, and glutamate cysteine ligase by broccoli seeds and isothiocyanates. J Nutr 134(12 Suppl):3499S–3506S

Morimitsu Y, Nakagawa Y, Hayashi K et al (2002) A sulforaphane analogue that potently activates the Nrf2-dependent detoxification pathway. J Biol Chem 277:3456–3463

Nakamura Y, Yoshida C, Murakami A et al (2004) Zerumbone, a tropical ginger sesquiterpene, activates phase II drug metabolizing enzymes. FEBS Lett 572:245–250

Nishinaka T, Ichijo Y, Ito M et al (2007) Curcumin activates human glutathione S-transferase P1 expression through antioxidant response. Toxicol Lett 170:238–247

Nguyen T, Nioi P, Pickett CB (2009) The Nrf2-antioxidant response element signaling pathway and its activation by oxidative stress. J Biol Chem 284(20):13291–13295

Prasad KN (2016) Simultaneous activation of Nrf2 and elevation of dietary and endogenous antioxidant chemicals for cancer prevention in humans. J Am Coll Nutr 35(2):175–184

Schadich E, Hlavac J, Volna T et al (2016) Effects of ginger phenylpropanoids and quercetin on Nrf2-ARE pathway in human BJ fibroblasts and HaCaT keratinocytes. Biomed Res Int 2016: 2173275

Saw CL, Kong TA (2011) Nuclear factor-erythroid 2-related factor 2 as a chemopreventive target in colorectal cancer. Expert Opin Ther Targets 15(3):281–295

Smith RE, Tran K, Smith CC et al (2016) The role of the Nrf2/ARE antioxidant system in preventing cardiovascular diseases. Diseases 4(4):34

Tan XL, Spivack SD (2009) Dietary chemoprevention strategies for induction of phase II xenobiotic-metabolizing enzymes in lung carcinogenesis: a review. Lung Cancer 65:129–137

Ulasov AV, Rosenkranz AA, Georgiev GP et al (2022) Nrf2/Keap1/ARE signaling: towards specific regulation. Life Sci 291:120111

de Vries HE, Witte M, Hondius D et al (2008) Nrf2-induced antioxidant protection: a promising target to counteract ROS-mediated damage in neurodegenerative disease? Free Radic Biol Med 45(10):1375–1383

14

Inducibility of Metabolizing Enzymes

An important toxicologically relevant feature associated with foreign compound metabolism is the inducibility of metabolizing enzymes. Some xenobiotics exhibit intrinsic toxicity, while some others are metabolically activated to potential toxicants by activation enzymes. The generated toxic metabolites then undergo detoxification enzyme–catalyzed reactions before they are ready for excretion.

Consequently, induction or inhibition of these metabolizing enzymes has a significant impact on the extent of toxicity of xenobiotics. Activation and detoxification enzymes have the potential to be inducted or inhibited by some chemical compounds. The inducibility of metabolic enzymes is of major significance, because it plays a role in the susceptibility of individuals to foreign compound–mediated toxicities.

Strategies for protecting cells from initiation toxic events include: a decrease in the expression of activation enzymes responsible for generating reactive species and an increase in the activities of detoxification enzymes that detoxify reactive chemical intermediates such as free radicals and electrophiles known to intervene cellular processes.

In either of the above events, it is important to realize that the consequence of enzyme induction or inhibition is dependent on not only the specific effect on certain activation enzymes or detoxification enzymes, but also the balance between the rates of activation and detoxification.

Induction of activation enzymes leads to a higher expression of enzyme activity, resulting in an enhancement in the activation rate and the production of potential toxicants. If the generated toxicants are not immediately removed from the body, the induction of activation enzymes would promote toxic effects of foreign compounds. In contrast, the induction of detoxification enzymes leads to a higher expression of enzyme activity and an increase in the rate of detoxifying reactions, resulting in a decrease in xenobiotic toxic effect and an acceleration of foreign compound excretion.

Moreover, the effects of inhibiting activation or detoxification enzymes on xenobiotic toxicity depend on not only the relative activities of these enzymes but also the nature of the foreign compounds. For a foreign compound whose toxicity is caused by metabolic activation, inhibition of the activation enzyme decreases xenobiotic toxic effects. While for a foreign compound with inherent toxicity, the inhibition of detoxification enzymes would increase xenobiotic toxicity. A further analysis of the effects of enzyme modulation is discussed below.

14.1 Inducibility of Activation Enzymes

When a foreign compound is metabolized to a potent intermediate, a significant induction of activation enzyme to a higher activity can increase the activation rate by producing more toxic intermediate, resulting in an enhancement in xenobiotic toxic effect. Hence, whether induction or inhibition of activation enzyme increases or decreases the toxicity of a foreign compound is an issue of complexity.

Among activation enzymes, CYP450 is most studied with respect to enzyme modulation.

A typical example of an inducer to CYP450 is phenobarbital, which has also been used for biochemical investigations of drug metabolism. The induction effect of phenobarbital mainly involves the activation of CYP450. Such induction was found to activate cocaine, leading to the potentiation of cocaine-induced hepatotoxicity.

Conversely, a modest inhibition of CYP450 to a lower activity can decrease the activation rate by producing less toxic intermediate, leading to a reduction in xenobiotic toxic effect. However, an over inhibition of activation rate can lead to an accumulation of foreign compounds and can cause impaired metabolic clearance. One typical inhibition example is the inhibition of CYP450 by grapefruit juice components. The consumption of grapefruit juice with drugs taken orally has been reported to inhibit intestinal CYP3A4 activity, resulting in a decrease in the metabolism of many drugs.

Figure 14.1 illustrates how induction and inhibition of a phase I activation enzyme may potentially affect xenobiotic toxicity. The figures reveal that the effects of activation enzyme modulation are dependent on both the resulted activation enzyme activity and the rate of respective detoxification enzyme.

Modest inhibition of activation enzyme may result in a lower xenobiotic toxic effect, assuming that the resulted enzyme activity is enough to perform its metabolic function. In contrast, a significant induction or inhibition of activation enzyme may enhance xenobiotic toxic effect, due to an increase in the rate of toxic intermediate production or an impairment in metabolic clearance.

14.2 Inducibility of Detoxification Enzymes

Among detoxification enzymes, UDP-glucuronosyl transferase (UGT), glutathione-S-transferases (GST), and NAD(P)H quinine reductase (NQO) are most studied with respect to enzyme modulation. In general, induction of phase II detoxification enzymes enhances the rate of detoxification reactions, leading to a faster increase

14.2 Inducibility of Detoxification Enzymes

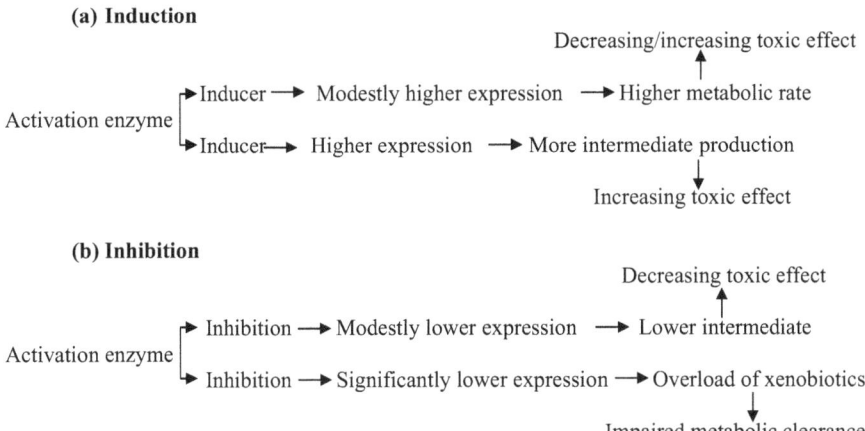

Fig. 14.1 Potential effects on xenobiotic toxicity by activation enzyme modulation

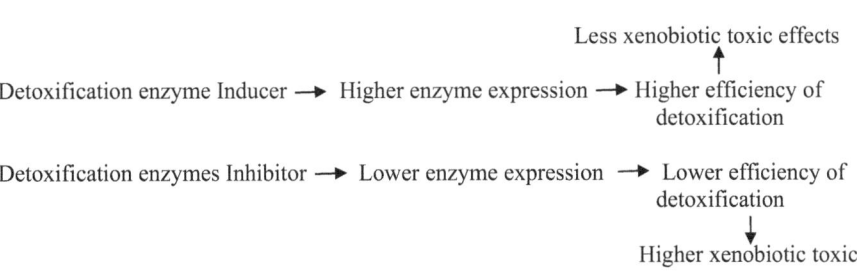

Fig. 14.2 Detoxification enzyme modulation affecting xenobiotic toxicity

in forming conjugated metabolites, and resulting in decreasing xenobiotic toxic effect and facilitating the excretion of xenobiotics. Conversely, inhibition of phase II detoxification enzymes decreases the rate of detoxification reactions, resulting in an increase in xenobiotic toxic effects.

Figure 14.2 illustrates how induction and inhibition of a detoxification enzyme affects xenobiotic toxicity. In comparison with Fig. 14.1, the modulation of detoxification enzyme is less complicated than that of activation enzyme.

As shown in Fig. 14.2, in general, the induction or inhibition of detoxification enzyme may lead to a lower or higher xenobiotic toxic effect, respectively. Nevertheless, when detoxification enzyme is overly induced, a potential problem involving the drug interaction and the action of other metabolisms may arise. Such issue requires further investigation.

An example of detoxification enzyme inducers is 1,2-dithiole-3-thione derivative, such as oltipraz. Oltipraz was reported to cause several-fold induction of constitutive hepatic and gastric activities of GST and NQO in mice. Such an increase in the expression of phase II enzyme genes is attributed to a reduction in aflatoxin B_1-induced hepatocarcinogenesis by oltipraz. Another example is

4-(Methylnitrosamino)-1-(3-pyridyl)-1-butanone (NNK), which is a tobacco-specific carcinogen. NNK exerts carcinogenic potential through hydroxylation in metabolic activation.

14.3 Lifestyle Modification

Two most known environment factors that are capable of modulating foreign compound–metabolizing enzymes are alcohol drinking and cigarette smoke. Alcohol drinking and cigarette smoke are tied to individual lifestyles. The metabolisms of foreign compounds are inducible by chronic alcohol consumption and cigarette smoking.

(a) Alcohol Drinking

An enhanced vulnerability of the alcoholic to the hepatotoxicity of many foreign compounds has been recognized. Activation enzyme CYP2E1 is the primary component of alcohol oxidizing system in the hepatic microsomes. CYP450 isoform plays a key role in alcoholic liver disease and associated oxidative stress.

CYP2E1-catalyzed reaction is also involved in the production of reactive oxygen species. Induction of CYP2E1 by alcohol increases metabolic activation of carcinogens. Moreover, heavy alcohol drinkers also develop severe hepatotoxicity from carbon tetrachloride exposure, but not for nondrinkers.

(b) Cigarette Smoke

Moreover, activation enzymes induced by cigarette smoking may enhance the metabolic activation of carcinogens. Such induction metabolism is a major mechanism involving the interactions between cigarette smoking and drugs. For example, polycyclic aromatic hydrocarbons, such as benz(a)pyrene and nitrosamines in cigarette smoke, undergo metabolic activation by phase I activation enzymes, particularly CYP450. Induction of CYP450 isozymes (CYP1A2 and CYP2E1) by cigarette smoking has been reported, where polycyclic aromatic hydrocarbons are believed to be responsible for the induction.

14.4 Monofunctional and Bifunctional Inducers

Two classes of enzyme inducers have been characterized: monofunctional and bifunctional inducers. Monofunctional inducers raise the activities of detoxification enzymes, including glutathione S-transferases, NAD(P)H:quinone reductase, UDP-glucuronosyl-transferases, in various tissues, while without significantly elevating the activities of activation enzymes such as CYP450 and flavor-containing monooxygenase. In contrast, bifunctional inducers raise the activities of both activation and detoxification enzymes.

Table 14.1 Examples of monofunctional and bifunctional Inducers

Monofunctional inducers	Bifunctional inducers
Phenols	Polycyclic aromatic hydrocarbons
Isothiocyanates	Flavonoids
Coumarins	Azo dyes
Thiocarbamates	
Cinnamates	
1,2-dithiol-3-thiones	

Table 14.1 lists some examples of monofunctional and bifunctional inducers. Reviewing the compounds listed in the table reveals that neither monofunctional nor bifunctional inducers display common characteristics in terms of chemical structures or functional groups.

For instance, quinone reductase carries out two-electron reductions to protect cells against the toxicity of quinones. This enzyme is induced in many tissues in coordination with other major phase II enzymes that protect against toxic intermediates. In quinine reductase activity assay, a monofunctional inducer elevates phase II enzyme in various tissues without significantly raising phase I activation enzyme such as CYP450 or aryl hydrocarbon hydroxylase.

In contrast, a bifunctional inducer induces both phase I and phase II metabolic enzymes. Bifunctional induction is dependent on aryl hydrocarbon receptor function or aryl hydrocarbon hydroxylase expression. For example, oltipraz has been recognized as a monofunctional inducer selectively activating phase II detoxification enzymes, such as glutathione S-transferase. However, oltipraz also activates xenobiotic responsive element from the phase I activation enzyme CYP1A1. Thus, Oltipraz is a bifunctional inducer, which modulates both phase I and II drug–metabolic enzymes.

Moreover, the induction potential of metanil yellow, a food color, on activation enzyme CYP450 and detoxification enzymes GST and QR was also investigated. Metanil yellow was found to cause significant induction of hepatic CYP450 and GST and QR activities. These results suggest that metanil yellow acts as a bifunctional inducer of metabolic enzymes.

14.5 Balance Between Activation and Detoxification Metabolisms

The efficiency of foreign compound metabolisms is dependent on the rates of activation and detoxification, which are interrelated to the activities of phase I enzymes and phase II enzymes. In general, minimal xenobiotic toxicity is found in either a fine balance rate between activation and detoxification or a higher rate of detoxification metabolism than activation metabolism.

Figure 14.3 summarizes the relative rates of activation and detoxification in relation to xenobiotic toxicity.

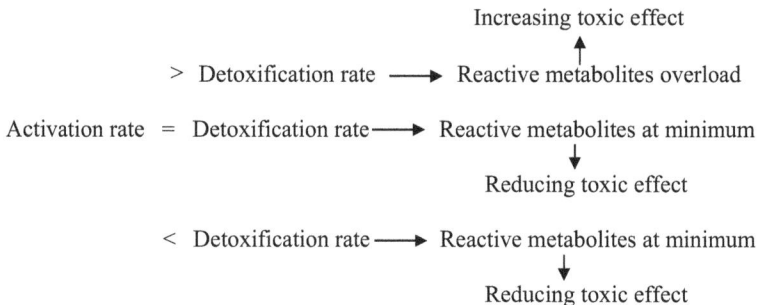

Fig. 14.3 Relative rates of activation and detoxification affecting xenobiotic toxicity

Induction of CYP450 expression is likely to either increase or decrease the toxic effects of foreign compounds, depending on the relative activity of phase II enzymes. When the induction of CYP450 results in the rate of activation comparable with that of detoxication, then the induction is likely to produce beneficial effects.

Nevertheless, if the induction of CYP450 results in the rate of activation being significantly higher than that of detoxification, then the induction may produce harmful effects, such as in the case of smoking-related induction of CYP450. Accordingly, the rate of reactive metabolite production comparable with the rate of detoxification reaction is essential to achieve this fine balance.

In general, detoxification metabolisms that function correctly appear to be an important means of preventing foreign compound–mediated toxic effects. Hence, a fine balance between activating enzymes and detoxification enzymes determines whether a toxic metabolite is detoxified or may cause cell dysfunction or damage.

14.6 Enzyme Modulation Against Potential Toxicity

Foreign compound–mediated toxic effects essentially occur in the following three circumstances: (1) xenobiotics are inherent toxicants that are detoxified by detoxification enzymes; (2) xenobiotics exhibit toxicity because of reactive intermediates formation in metabolism catalyzed by activation enzymes, which are not effectively detoxified by detoxification enzymes; and (3) xenobiotic toxic effects are potentiated by other foreign compounds, which may have little or no toxicity but are capable of interfering effectively with detoxification enzymes.

In circumstances (1) and (3), the induction of detoxification enzymes is beneficial to protect against potential foreign compound–mediated toxic effects. While in circumstance (2), either the inhibition of activation enzymes or the induction of detoxification enzymes may be helpful in achieving such a protection.

14.6.1 Enzyme Modulation

It has been proposed that modulation of foreign compound–metabolizing enzymes may be a useful approach in minimizing the risk of xenobiotics-mediated toxic effects. One hypothesis proposes to protect against xenobiotic toxic effects by inhibiting activation enzymes as well as inducing detoxification enzymes. Another hypothesis proposes that the induction of detoxification enzymes alone is enough to achieve the protection against carcinogenesis and other forms of toxicities. In general, the inhibition of activation enzymes to affect xenobiotic toxicity is more complicated than the induction of phase II detoxification enzymes.

14.6.2 Hypothesis of Detoxification Enzyme Induction

Extensive studies have been carried out to test the hypothesis that the induction of detoxification enzymes alone is enough to achieve the protection against xenobiotic toxicity. Results of these investigations suggested that the induction of phase II detoxification enzymes appears to be an effective means for achieving protection against a variety of carcinogens. In line with these studies, it has been proposed that a promising approach to minimize xenobiotic toxic effects and related oxidative stress is to increase the intake of phase II enzyme inducers.

Besides this hypothesis, it has also been postulated that the ingestion of an excessive amount of antioxidant is presumed to shift the oxidant–antioxidant balance toward the antioxidant side. Although an excessive antioxidant may effectively break down reactive oxygen species including free radicals, however, such a shift of balance towards the antioxidant side by taking excessive amounts of antioxidants could result in affecting some key physiological processes that are dependent on free radicals. Such issue is a matter of concern and requires further investigations.

The subjects of induction and inhibition compounds of importance in modulating xenobiotic-metabolic enzymes and dietary inducers of phase II detoxification enzymes are further discussed in another chapters.

14.7 Inducer–Metabolic Enzyme Interactions

By interacting with metabolic enzyme covalently or noncovalently, an inducer molecule can affect the activity of enzyme by the following two manners: (1) modifying the conformation of the enzyme or (2) affecting reactive intermediate generation through inducer–metabolic enzyme interaction. As a result, the activity of enzyme is either increased or decreased, depending on whether the inducer leads to more or less favorable enzyme conformation or interaction for catalytic action during xenobiotic metabolism.

(a) Modification of Enzyme Conformation

Elucidation of a potential change in enzyme conformation owing to inducer–enzyme interaction is essential to an understanding of underlying factors involving induction or inhibition of metabolic enzymes mediated by an inducer. In studying how inducers affecting the interaction of foreign compounds with metabolic enzymes, research has suggested that inducers covalently react with cellular molecules by virtue of highly reactive functional group (such as sulfhydryl). Nevertheless, related research involving enzyme conformation with respect to inducer–metabolic enzyme interaction requires future investigations.

(b) Enzyme Interaction for Catalytic Action

To exert their toxic effects, most foreign compounds require activation catalyzed by activation enzymes, mostly cytochrome P450, to form electrophilic reactive intermediates or metabolites. Such reactive groups of intermediate or metabolite are substrates of metabolic enzymes. They are essentially electrophiles with either positive charge or carrying a partial positive charge. Such electrophilic reactive intermediate or metabolite prefers to interact with a nucleophilic group in the amino acid side chain of metabolic enzymes, such as serine and tyrosine hydroxyl, aspartate or glutamate carboxylase, histidine imidazole, and cysteine sulfhydryl groups.

Moreover, the substrate for detoxification enzyme conjugation reaction often shares a common feature that contains an electrophilic atom. For instance, in detoxification enzyme–catalyzed conjugation reaction, the site of glucuronidation generally involve an electron–rich nucleophilic O, N or S atom. Such glucuronidation is the primary metabolic reaction for many compounds containing functional groups such as -OH, -COOH, -SH, and -NH2.

14.7.1 Michael Reaction Acceptors

As discussed above, induction of detoxification enzymes is a major strategy for reducing the susceptibility of living cells to the toxic and carcinogenic effects caused by metabolic reactive intermediates, including reactive oxygen species and free radicals. A diversity of small molecules of naturally occurring or synthetic origins are capable of inducing detoxication enzymes. Evaluation of their chemical structures reveals that detoxification enzyme inducers belong to a variety of chemical classes that have the ability to modify sulfhydryl group of a cysteine residue in metabolic enzymes.

Moreover, extensive research in inducer–enzyme interactions has also revealed that many detoxication enzyme inducers contain Michael acceptor functionalities, such as in olefins or acetylenes. The potency of inducers was also found to parallel with their reactivity as Michael acceptors. For instance, olefins, unsaturated

14.7 Inducer–Metabolic Enzyme Interactions

Table 14.2 Metabolic enzyme inducers with Michael reaction acceptors

Classes	Typical compounds
Sulforaphane and isothiocycanate	Sulforaphane; Phenylethyl isothiocyanate
1,2-dithiole-3-thione and derivatives	1,2-dithioe-3-thione;
	4-methyl-5-pyrazinyl-D3T
Indole-3-carbinol	Indole-3-carbinol
Flavonoids	Catechin
Epicatechin	Epigallocatechin; Leucocyanidines
Quercetin	Myricetin; Fistein
Isoliquiritigenin	Diosmin; Hesperidin
Isoflavones	Genistein
Daidzein	Isoflavone
Phenols and polyphenols	Resveratrol; Curcumin; Gallic acid
	Rosmarinic acid; Carnosic acid
Ellagic acid	Tannin
Protocatechuic acid	
Organosulfur	Diallyl sulfide; diallyl trisulfide; alliin
Terpenes and terpenoids	Beta-carotene; lycopene
Quinoline	Ethoxyquinn
Canthaxanthin	
Astaxanthin	
Zerumbone	
Limonene	
Others	Nivalenol

hydrocarbons, are made up of hydrogen and carbon that contain one or more pairs of carbon–carbon double bond (C=C).

For example, acetylenes are chemical compounds that consist of the formula of HC ≡ CH (carbon–carbon triple bond). Michael acceptors are capable of conjugation with electron-withdrawing groups, such as carbonyl [C=O], sulfinyl [S=O], nitrile [C ≡ N], or acetylene [C ≡ C]. Table 14.2 lists detoxification enzyme inducers with Michael acceptor functionalities, where the typical functional groups of Michael reaction receptors are alpha, beta-unsaturated aldehydes, ketones, quinones, thioketones, sulfones, esters, nitriles, and nitro groups.

14.7.2 Unsaturated Carbon–Carbon Bonds

Many of inducers contain Michael reaction acceptors, such as olefins or acetylenes. Olefins are unsaturated hydrocarbons that contain one or more pairs of carbon atoms linked by a double bond. Acetylenes are also unsaturated hydrocarbons but consist of two carbon atoms bonded together by a triple bond.

The carbon–carbon double or triple bond serves as a source of electrons, which availability is determined by the groups attached to it. An electron-withdrawing group attached to the carbon–carbon double or triple bond (such as C=O, -COOH, -COOR, or -CN) is capable of activating the double or triple bond toward the reagents that are rich in electron.

14.7.3 Phenolic Hydroxyl Groups

Studies of metabolic enzyme inducers revealed that the presence of hydroxyl groups at ortho- position on the aromatic rings is an important structural element for high inducer potency. Ortho-hydroxyl groups have a significant impact on the protective role of detoxification enzyme inducers. Table 14.3 presents some inducers that contain an ortho-hydroxyl group on the aromatic ring.

Inducers such as flavonoid and curcuminoid analogues contain phenolic hydroxyl groups in addition to Michael acceptor functionalities. Since phenol hydroxyl groups can scavenge oxygen- and nitrogen-centered reactive intermediates. Flavonoids and curcuminoids play not only an indirect protective role by inducing detoxification enzymes but also a direct protective role by scavenging hazardous oxidants.

Moreover, the introduction of ortho-hydroxyl groups to the aromatic rings of phenyl propenoids was found to enhance their potencies not only as inducers of quinone reductase, but also as quenchers of superoxide. The presence of an ortho-hydroxyl substituent on the aromatic ring was also found to profoundly increase the induction potency of benzylidene-alkenone on quinone reductase activity, which also appears to be correlated with their ability to quench superoxide radicals.

Table 14.3 Metabolic enzymes inducers with ortho-hydroxyl group on aromatic ring

Class of compound	Chemical modulator
Flavonoids	Catechin
Epicatechin	Epigallocatechin, Leucocyanidines
Quercetin	Myricetin, Fisetin
Hesperidin	Diosmin
Phenols and polyphenols (Gallic, rosmarinic, carnosic, ellagic, and protocatechuic acids)	Curcumin, Tannin

Bibliography

Albena T, Dinkova-Kostova MA et al (2001) Potency of Michael reaction acceptors as inducers of enzymes that protect against carcinogenesis depends on their reactivity with sulfhydryl groups. Proc Natl Acad Sci 98(6):3404–3409

Buetler TM, Gallagher EP, Wang C et al (1995) Induction of phase I and phase II drug-metabolizing enzyme mRNA, protein, and activity by BHA, ethoxyquin, and oltipraz. Toxicol Appl Pharmacol 135:45–57

Chen C-H (2020) Xenobiotic metabolic enzymes: bioactivation and antioxidant defense. Springer Nature, Cham

Chen Y, Huang C, Zhou T et al (2008) Genistein induction of human sulfotransferases in HepG2 and Caco-2 cells. Basic Clin Pharmacol Toxicol 103(6):553–559

Cuendet M, Oteham CP, Moon RC et al (2006) Quinone reductase induction as a biomarker for cancer chemoprevention. J Nat Prod 69:460–463

Das M, Ramchandani S, Upreti RK et al (1997) Metanil yellow: a bifunctional inducer of hepatic phase I and phase II xenobiotic-metabolizing enzymes. Food Chem Toxicol 35(8):835–838

De Long MJ, Dolan P, Santamaria AB et al (1986) 1,2-Dithiol-3-thione analogs: effects on NAD (P)H:quinone reductase and glutathione levels in murine hepatoma cells. Carcinogenesis 7(6): 977–980

De Long MJ, Santamaria AB, Talalay P (1987) Role of cytochrome P1-450 in the induction of NAD(P)H:quinone reductase in a murine hepatoma cell line and its mutants. Carcinogenesis 8(10):1549–1553

Debersac P, Heydel JM, Amiot MJ et al (2001) Induction of cytochrome P450 and/or detoxication enzymes by various extracts of rosemary: description of specific patterns. Food Chem Toxicol 39(9):907–918

Dinkova-Kostova AT, Holtzclaw WD et al (2002) Direct evidence that sulfhydryl groups of Keap1 are the sensors regulating induction of phase 2 enzymes that protect against carcinogens and oxidants. Proc Natl Acad Sci 99(18):11908–11913

Dinkova-Kostova AT, Fahey JW, Talalay P (2004) Chemical structures of inducers of nicotinamide quinone oxidoreductase 1 (NQO1). Methods Enzymol 382:423–448

Fong AT, Swanson HI, Dashwood RH et al (1990) Mechanisms of anti-carcinogenesis by indole-3-carbinol. Studies of enzyme induction, electrophile-scavenging, and inhibition of aflatoxin B1 activation. Biochem Pharmacol 39:19–26

Jakel RJ, Townsend JA, Kraft AD et al (2007) Nrf2-mediated protection against 6-hydroxydopamine. Brain Res 1144:192–201

Kensler TW, Curphey TJ, Maxiutenko Y et al (2000) Chemoprotection by organosulfur inducers of phase 2 enzymes: dithiolethiones and dithiins. Drug Metabol Drug Interact 17(1–4):3–22

Liu Y, Kern JT, Walker JR et al (2007) A genomic screen for activators of the antioxidant response element. Proc Natl Acad Sci USA 104:5205–5210

Lnenickova K, Skalova L, Stuchlikova L et al (2018) Induction of xenobiotic-metabolizing enzymes in hepatocytes by beta-naphthoflavone: time-dependent changes in activities, protein and mRNA levels. Acta Pharma 68(1):75–85

Miao W, Hu L, Kandouz M et al (2003) Oltipraz is a bifunctional inducer activating both phase I and phase II drug-metabolizing enzymes via the xenobiotic responsive element. Mol Pharmacol 64(2):346–354

Murphy SE, Nunes MG, Hatala MA (1997) Effects of phenobarbital and 3-methylcholanthrene induction on the formation of three glucuronide metabolites of 4-(methylnitrosamino)-1-(3-pyridyl)-1-butanone, NNK. Chem Biol Interact 103:153–166

Nguyen T, Sherratt PJ, Pickett CB (2002) Regulatory mechanisms controlling gene expression mediated by the antioxidant response element. Annu Rev Pharmacol Toxicol 43:233–260

Okey AB, Roberts EA, Harper PA et al (1986) Induction of drug-metabolizing enzymes: mechanisms and consequences. Clin Biochem 19:132–141

Prestera T, Talalay P (1995) Electrophile and antioxidant regulation of enzymes that detoxify carcinogens. Proc Natl Acad Sci USA 92(19):8965–8969

Prochaska HJ, Talalay P (1998) Regulatory mechanisms of monofunctional and bifunctional anticarcinogenic enzyme inducers in murine liver. Cancer Res 48:4776–4782

Prochaska HJ, Santamaria AB, Talalay P (1992) Rapid detection of inducers of enzymes that protect against carcinogens. Proc Natl Acad Sci USA 89:2394–2398

Ramos-Gomez M, Kwak MK, Dolan PM et al (2001) Sensitivity to carcinogenesis is increased and chemoprotective efficacy of enzyme inducers is lost in nrf2 transcription factor-deficient mice. Proc Natl Acad Sci USA 98:3410–3415

Rushmore TH, Pickett CB (1990) Transcriptional regulation of the rat glutathione S-transferase Ya subunit gene. Characterization of a xenobiotic-responsive element controlling inducible expression by phenolic antioxidants. J Biol Chem 265(24):14648–14653

Shen G, Kong AN (2009) Nrf2 plays an important role in coordinated regulation of Phase II drug metabolism enzymes and Phase III drug transporters. Biopharm Drug Dispos 30:345–355

Talalay P (1989) Mechanisms of induction of enzymes that protect against chemical carcinogenesis. Adv Enzym Regul 28:237–250

Talalay P, Fahey JW, Holtzclaw WD et al (1995) Chemoprotection against cancer by phase 2 enzyme induction. Toxicol Lett 82-83:173–179

Ushida Y, Talalay P (2013) Sulforaphane accelerates acetaldehyde metabolism by inducing aldehyde dehydrogenases: relevance to ethanol intolerance. Alcohol Alcohol 48(5):526–534

Yamaori S, Kinugasa Y, Jiang R et al (2015) Cannabidiol induces expression of human cytochrome P450 1A1 that is possibly mediated through aryl hydrocarbon receptor signaling in HepG2 cells. Life Sci 136:87–93

Yang CS, Chhabra SK, Hong JY et al (2001) Mechanisms of inhibition of chemical toxicity and carcinogenesis by diallyl sulfide (DAS) and related compounds from garlic. J Nutr 131:1041S–1045S

Yannai S, Day AJ, Williamson G et al (1998) Characterization of flavonoids as monofunctional or bifunctional inducers of quinone reductase in murine hepatoma cell lines. Food Chem Toxicol 36(8):623–630

Zevin S, Benowitz NL (1999) Drug interactions with tobacco smoking. An update. Clin Pharmacokinet 36:425–438

Inducibility of Metabolizing Enzymes

15

Metabolic enzyme inducibility is of major significance, because it plays an important role in the susceptibility of individuals to foreign compound–mediated toxicities. Although some xenobiotics exhibit intrinsic toxicity, while some others are metabolically activated to potential toxicants by activation enzymes. The generated metabolites undergo detoxification enzyme–catalyzed reactions for detoxification, before foreign compounds are ready for excretion.

Induction or inhibition of metabolic activation and detoxification enzymes has a significant impact on the extent of toxicities of xenobiotics. The effects of metabolic enzymes on xenobiotic toxicity depend on not only the nature of the foreign compounds but also the relative activities of these enzymes. If the generated toxicants are not immediately removed from the body, the induction of activation enzymes would promote toxic effects of foreign compounds.

The expression of metabolic enzymes can also differ significantly as a result of exposure to environmental chemicals or lifestyle differences.

In addition, metabolic enzymes have the potential to be inducted or inhibited by some chemical compounds. Strategies for protecting cells from initiation toxic events include (a) a decrease in the expression of activation enzymes responsible for the generation of reactive species and (b) an increase in the activities of detoxification enzymes that detoxify reactive intermediate species that intervene cellular processes.

In either case, it is important to realize that the consequence of enzyme induction or inhibition is dependent on not only the specific effect on certain activation or detoxification enzymes, but also the balance between the rates of activation and detoxification.

15.1 Defense Against Potential Metabolic Toxicity

Induction of activation enzymes can lead to a higher expression of enzyme activity, resulting in an enhancement in the production of potential toxic metabolic reactive intermediates. Hence, for a foreign compound whose toxicity is caused by metabolic activation, inhibition of activation enzymes will decrease toxic effects of foreign compounds. However, an over inhibition of activation rate can lead to an accumulation of foreign compounds and, therefore, can cause impaired metabolic clearance.

15.1.1 Modification of Activation Enzymes

CYP450, the major activation enzyme responsible for a majority of activation reactions, involves activation of various foreign compounds, including chemicals, drugs, and pesticides. The modulation of CYP450 activity as a defense against xenobiotic-mediated toxic effects is complicated, because the result of modulation is related to the respective activity of detoxification enzymes. An induction of CYP450 could either decrease or increase xenobiotic toxic effects, depending on whether detoxification activity is comparable with the induced CYP450 activity or not.

If the rate of activation is higher than that of detoxification, an overload of potentially toxic intermediates occurs, leading to an increase in toxic effects. In this circumstance, an inhibition of CYP450 could cause either a decrease in xenobiotics-mediated toxic effects due to producing less chemically activated intermediates or an increase in toxicities due to the accumulation of xenobiotics and impaired metabolic clearance.

15.1.2 Modification of Detoxification Enzymes

In contrast, the induction of detoxification enzymes can lead to a higher expression of enzyme activities and an increase in the rate of detoxifying reactions. These can result in a decrease in xenobiotic toxic effect and an acceleration of foreign compound excretion. On the other hand, for a foreign compound with inherent toxicity, the inhibition of detoxification enzymes would increase xenobiotic toxicity.

Moreover, chemical compounds that modulate the expression of phase II enzymes can play an important role in the intervention of toxicity processes, particularly at the initial stage where activated reactive intermediates are involved.

Considerable evidences have been accumulated to support the proposal that the induction of phase II detoxification enzymes is a useful approach for defense against the risk of exposure to xenobiotic toxicity. Such observation further supports the proposal that the induction of detoxification enzymes is a potential defense mechanism of chemoprevention.

15.1.3 Antioxidant Activities

Glutathione combining with its coupled glutathione S-transferase enzyme system is one of the most important antioxidant defense mechanisms in the body. The reactive sulfhydryl group of glutathione is responsible for glutathione antioxidant activities. A high level of intracellular glutathione plays a prominent role in antioxidant protection against reactive oxygen species.

However, a reduction of the amount of glutathione to a lower level can impair the defense of the cells against toxic effects. The depletion of glutathione can occur due to a lack of essential nutrients or amino acids needed to synthesize glutathione. Lifestyle effects such as smoking can deplete the level of glutathione.

15.2 Modification of Metabolic Enzymes

As discussed above, when a foreign compound is metabolized to a reactive intermediate, a significant induction of activation enzyme to a higher activity will result in increasing the activation rates and producing more toxic intermediates, leading to a potential enhancement in xenobiotic toxic effect. On the other hand, a significant inhibition of activation enzymes may cause an impairment in metabolic clearance and xenobiotic toxic effect.

However, an induction of phase II detoxification enzymes can enhance the rate of detoxification reactions, leading to a faster increase in forming conjugated or nonconjugated metabolites, which result in decreasing xenobiotic toxic effect and facilitating the excretion of xenobiotics. Conversely, inhibition of phase II detoxification enzymes decreases the rate of detoxification reactions, which can result in an increase in xenobiotic toxic effects.

15.2.1 Modification of Activation Enzymes

Among activation enzymes, CYP450 is most studied with respect to enzyme modulation.

The induction of CYP450 affects its activation. A typical example of CYP450 inducer is phenobarbital, which has been used for biochemical investigations of drug metabolism. Phenobarbital was reported to induce drug metabolism in both animal models and humans. Phenobarbital was also found to enhance the activation of cocaine, leading to the potentiation of cocaine-induced hepatotoxicity.

Another example of inducer is 4-(Methylnitrosamino)-1-(3-pyridyl)-1-butanone (NNK), which is a tobacco-specific carcinogen. NNK exerts carcinogenic potential through hydroxylation reaction in metabolic activation. If the resulted modification is enough to perform activation metabolic function, a modest inhibition of CYP450 can lead to a lower activation rate and produce less toxic intermediate. This can result in a reduction in xenobiotic toxic effect.

(a) Induction

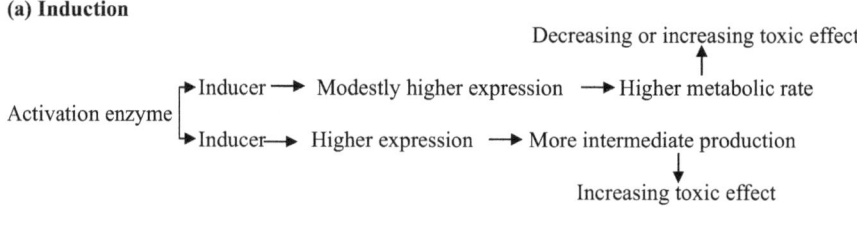

(b) Inhibition

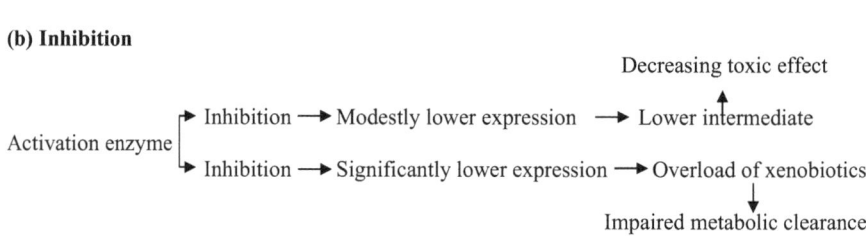

Fig. 15.1 Potential effects on xenobiotic toxicity by activation enzyme modulation

However, it should be noted that the effects of activation enzyme modulation are not dependent on the resulted activation enzyme activity alone. The rate of respective detoxification enzyme is also an important factor.

Figure 15.1 illustrates how induction and inhibition of a phase I activation enzyme may potentially affect foreign compound toxicity. The figure indicates that if the resulted enzyme activity is enough to perform its metabolic function, a modest inhibition of activation enzyme may result in a lower xenobiotic toxic effect. However, the effects of activation enzyme modulation are dependent on not only the resulted activation enzyme activity but also the rate of respective detoxification enzyme.

15.2.2 Modulation of Detoxification Enzymes

Among detoxification enzymes, UDP-glucuronosyl transferase (UGT), glutathione-S-transferases (GST), and NAD(P)H quinine reductase (NQO) are most studied with respect to their enzyme modulation. Generally, induction of phase II detoxification enzymes enhances the rate of detoxification reactions, leading to a faster increase in forming conjugated or nonconjugated metabolites, which results in decreasing xenobiotic toxic effect and facilitating the excretion of xenobiotics. Conversely, the inhibition of phase II detoxification enzymes decreases the rate of detoxification reactions, resulting in an increase in xenobiotic toxic effects.

(a) Potential Modulation Effects

Figure 15.2 illustrates how induction and inhibition of phase II detoxification enzymes may potentially affect xenobiotic toxicity. The figure reveals that

15.3 Major Inducers of Metabolic Enzymes

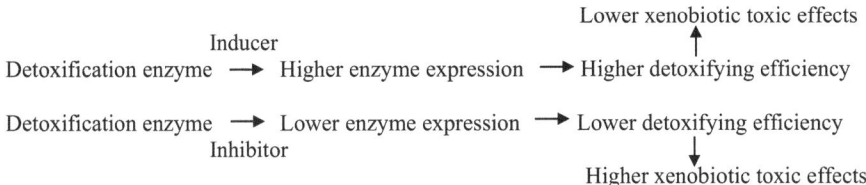

Fig. 15.2 Potential effects on xenobiotic toxicity by detoxification enzyme modulation

detoxification enzyme modulation are dependent on the rate of respective detoxification enzyme.

For example, inducer oltipraz, a 1,2-dithiole-3-thione derivative, was reported to cause inductions of constitutive hepatic activities of GST and NQO in mice. Such increases in the expression of phase II genes are attributed to a reduction of aflatoxin B_1-induced hepatocarcinogenesis by oltipraz.

(b) Hypothesis Underlying Defense Mechanism

Activation and detoxification enzymes protect the cells from toxic effects mediated by a variety of foreign compounds. Such hypotheses underlying the defense against xenobiotics-mediated toxicity include: (a) the reduction of activation by inhibiting phase I activation enzymes and (b) the enhancement of detoxification by inducing phase II detoxification enzymes.

Hence, induction of detoxification enzyme may result in a lower xenobiotic toxic effect. Conversely, Inhibition of detoxification enzyme may lead to higher xenobiotic toxic effect.

15.3 Major Inducers of Metabolic Enzymes

Known compounds of importance in the modulation of foreign compound metabolic enzymes include: (a) sulforaphane and other isothiocyanates, (b) 1,2-dithiole-3-thione and derivatives, (c) indole-3-carbinol, (d) flavonoids and isoflavones, (e) polyphenols, (f) organosulfur compounds, and (g) terpenes and terpenoids.

Other induction compounds that are not discussed here include geniposide. Geniposide was reported to inhibit liver CYP450-dependent monooxygenases, increase hepatic glutathione, and induce glutathione S-transferase activity in the liver.

Among these chemical compounds, oltipraz (1,2-dithiole-3-thione derivative) and sulforaphane (4-methyl-sulfinylbutane isothiocyanate) are the most studied inducers of detoxification enzymes. These major metabolic enzyme inducers and their effects on metabolic enzyme modulation are discussed below.

15.3.1 Sulforaphane and Other Isothiocyanates

Cruciferous vegetables contain a variety of glucosinolates that form different isothiocyanates under hydrolysis reactions. Glucoraphanin is the precursor of sulforaphane. Sulforaphane is produced from glucoraphanin by myrosinase, a class of enzymes that catalyzes the hydrolysis of glucosinolates.

Other isothiocyanates of interest include allyl, phenethyl, and benzyl isothiocyanates.

Isothiocyanates contain –N=C=S group formed by substituting sulfur of isocyanates for oxygen. They are hydrolysis products of the enzymatic conversion of metabolites called glucosinolates (sulfur-containing compounds).

Isothiocyanates are primarily metabolized through the mercapturic acid pathway, giving rise to N-acetylcysteine conjugates. Many isothiocyanates, particularly sulforaphane, are potent inducers of detoxification enzymes, including glutathione S-transferases, UDP-glucuronosyl transferases, and quinone reductase. Moreover, anticarcinogenic effects of isothiocyanates also appear to be mediated by the suppression of carcinogen activation by CYP450 isozymes through inhibition and regulation of their catalytic activities.

Sulforaphane has been reported to significantly induce detoxification enzyme activity in human prostate cells. Oral administration of sulforaphane also potently induces phase II enzymes in the bladder tissues. Induction of glutathione S-transferases and quinone reductases by sulforaphane was reported to block diesel exhaust particles–induced enhancement of immunoglobulin IgE production. Diesel exhaust particles initiate and intensify airway allergic responses through enhancing IgE production. These exhaust particles are associated with allergic respiratory disorders including asthma.

Inhibition of NNK-induced lung tumorigenesis by phenetyl isothiocyanate has been reported to block the metabolic activation of nitrosamine ketone (NNK). CYP450 2B1 is one of the major isozymes involved in the activation of nicotine-derived NNK. NNK is the most potent carcinogen present in tobacco. Phenetyl isothiocyanate is also an effective inhibitor of lung tumor induction by the tobacco-specific nitrosamine, 4-(methylnitrosamine)-1-(3-pyridyl)-1-butanone.

Table 15.1 shows that sulforaphane and other isothiocyanates act as not only the inducers for detoxification enzymes but also the inhibitors for activation enzymes. A number of isothiocyanates, including sulforaphane, have been reported to modulate activation and detoxification enzymes.

15.3.2 1,2-Dithiole-3-Thione and Derivatives

1,2-dithiole-3-thiones (D3Ts) are naturally occurring five-membered cyclic organosulfur compounds. Some derivatives of D3T have been shown to induce phase II detoxification enzymes and to protect against chemical carcinogenesis. The most extensively studied D3T derivative is oltipraz (4-methyl-5-pyrazinyl-3H-

15.3 Major Inducers of Metabolic Enzymes

Table 15.1 Enzyme modulation by isothiocyanates

a. Inhibition of activation enzymes	
Compound	Enzyme inhibition
Sulforaphane	CYP450
Phenylpropylisothiocyanate	CYP450
Phenylhexylisothiocyanate	CYP450
Benzylisothiocyanate	CYP450
Phenylethylisothiocyanate	CYP450
Phenylisothiocyanate	CYP450
b. Induction of detoxification enzymes	
Compound	Enzyme induction
Sulforaphane	GST, QOR, epoxide hydrolase
Phenylethylisothiocyanate	GST, QOR
Allyisothiocyanate	GST
Benzylisothiocyanate	GST

Table 15.2 Enzyme modulation by 1,2-dithiole-3-thione and derivatives

a. Inhibition of activation enzymes	
Compound	Enzyme inhibition
1,2-dithiole-3-thione	CYP450[3]
Oltipraz	CYP450
b. Induction of detoxification enzymes	
Compound	Enzyme induction
1,2-dithiole-3-thione	GST
Oltipraz	QOR, GST, UGT
5,6-dihydrocyclopenta-1,2-dithiole-3-thione	QOR, GST
4-chloro-5-methyl-1,2-dithiole-3-thione	QOR, GST
4-phenyl-1,2-dithiole-3-thione	QOR, GST

1,2-dithiole-3-thione). Its chemoprotective action may offer protection against a wide range of carcinogens.

Rodent model studies have demonstrated chemoprevention of aflatoxin-induced hepatocarcinogenesis by oltipraz. Administration of oltipraz was reported to protect mice against the neoplasia induced by benzo[a]pyrene. The chemoprotective effect of oltipraz is attributed, in part, to the induction of detoxification enzymes, such as glutathione S-transferases, UDP-glucuronosyl transferases and quinone oxidoreductases in the liver and other target tissues.

The increased expression of detoxification enzymes is of central importance to chemoprevention. The inhibition of activation enzymes by CYP450 isozymes may also attributes to oltipraz effect. Several other D3T derivatives, such as 5,6-dihydrocyclopenta-D3T, 4-chloro-5-methyl-D3T and 4-phenyl-D3T, were found to exhibit protection against acute toxicity of many foreign compounds and offer effective inhibition of carcinogenesis.

Table 15.2 lists D3T, oltipraz, and other D3T derivatives that act as inhibitors of activation enzymes or inducers of detoxification enzymes.

15.3.3 Indole-3-Carbinol

Indole-3-carbinol is the hydrolysis product of glucobrassicin, which is found at relatively high levels in cruciferous vegetables. When plant cells are damaged by chopping or chewing, glucobrassicin interacts with myrosinase, resulting in the formation of indole-3-carbinol.

Myrosinase, an enzyme that catalyzes the hydrolysis of glucosinolates, is physically separated from glucosinolates in intact plant cells.

Oral consumption of indole-3-carbinol leads to the formation of acid condensation products such as dimeric 3,3'-diindolylmethane, which are responsible for the biological effects of indole-3-carbinol. As a potential chemoprevention agent, indole-3-carbinol is a compound of growing interest. Indole-3-carbinol was found to induce hepatic levels of CYP1A1 but inhibit flavin-containing monooxygenase in rat livers and intestines.

Studies of CCl_4-induced hepatotoxicity revealed that indole-3-carbinol induces the level of CYP450 activity. However, the produced decrease in its activity by CCl_4 is restored to the control level by indole-3-carbinol. Indole-3-carbinol was also found to inhibit aflatoxin-induced hepatocarcinogenesis in rats.

Reports of indole-3-carbinol on the modulation of activation enzymes (such as CYP450 and FMO) and detoxification enzymes (such as GST and QOR) are presented in Table 15.3. The table reveals that indole-3-carbinol affects both phase I activation enzymes and phase II detoxification enzymes.

15.3.4 Flavonoids and Isoflavones

Flavonoids are the most common group of polyphenolic compounds that are synthesized by plants. Based on their chemical structures, flavonoids are grouped into flavonols, flavones, flavanones, isoflavones, catechins, anthocyanidins, and chalcones. Many biological effects of flavonoids appear to be associated with their ability to modulate cell-signaling pathways. However, a number of studies have also revealed the effects of flavonoids on foreign compound-metabolic enzymes.

Table 15.3 Enzyme modulation by indole-3-carbinol

a. Inhibition of activation enzymes	
Compound induction	Enzyme inhibition
Indole-3-carbinol	CYP450
Indole-3-carbinol	FMO
b. Induction of detoxification enzymes	
Compound	Enzyme induction
Indole-3-carbinol	GST, QOR

15.3 Major Inducers of Metabolic Enzymes

Table 15.4 Enzyme modulation by flavonoids

a. Inhibition/induction of activation enzymes

Compounds	Inhibition	Induction	Unaffected
4′-bromoflavone	CYP450		CYP450
Beta-naphthoflavone		CYP450	
Leucocyanidines	CYP450		
Catechin	CYP450		
Epigallocatechin		CYP450	
Quercetin			CYP450

b. Induction of detoxification enzymes

Compound	Enzyme induction
Isoliquiritigenin	QOR
4′-bromoflavone	QOR, GST
Beta-naphthoflavone	GST, UGT, NQO
Leucocyanidines	GST
Anthocyanins	GTS
Catechin	UGT
Quercetin	QOR, QR
Myricetin	QOR

(a) Flavonoids

The protective effects of flavonoids against xenobiotic toxicity are attributed in part to the modulation of metabolic enzymes. For example, leucocyanidine is known to exhibit antioxidant and antimutagenic activities, and exert a protective effect against cardiovascular disease. Its role as a chemoprevention agent against toxic or carcinogenic metabolites involves the induction of detoxification enzymes as well as the inhibition of activation enzymes.

It has been proposed that modulation of detoxification enzymes, such as uridine-diphosphate-glucuronosyltransferase, glutathione S-transferase, and quinone oxidoreductase, to accelerate detoxification of carcinogens is an important mechanism of the anticarcinogenic effects of flavonoides. For example, isoliquiritigenin is an inducer of quinone reductase and 4′-bromoflavone. It significantly induces quinone reductase in addition to glutathione S-transferase.

Table 15.4 presents metabolic enzyme modulation by flavonoids. The table lists a number of flavonoids that are reported to modulate detoxification enzymes and activation enzymes. Although flavonoids induce detoxification enzymes, their effects on the activation enzyme CYP450 are not homogenous, which can be induction, inhibition, or no significant effect.

(b) Isoflavones

Isoflavones comprise a class of organic compounds related to flavonoids. Soybeans and soy products are rich sources of isoflavones in the human diet. Genistein and daidzein are two of several known isoflavones found in plants and

Table 15.5 Enzyme modulation by isoflavones

a. Inhibition/induction of activation enzymes		
Compound	Enzyme inhibition	Enzyme induction
Genistein	CYP450	
Soy isoflavones		CYP450
b. Induction of detoxification enzymes		
Compound	Enzyme induction	
Genistein	QR	
Daidzein	QR	
Soy Isoflavones	GST, QR and UGT	

herbs. Isoflavones were reported to regulate the expression of genes critical to drug metabolism.

When rats consume a diet high in isoflavones, the activities of glutathione S-transferase in the kidney and quinone reductase in the colon are higher. Genistein was found to induce the activity of quinone reductase and inhibit the expression of aromatase. The inhibition of aromatase leads to a decrease in estrogen biosynthesis, thus producing an antiestrogenic effect.

Table 15.5 lists some isoflavones that are capable of modulating detoxification enzymes and activation enzymes. In addition to inducing detoxification enzymes, genistein and soy isoflavones can also affect the activity of CYP450.

15.3.5 Polyphenols

Polyphenols are characterized by the presence of more than one phenol group in their chemical structures. Polyphenolic compounds are a group of chemical compounds present in beverages, such as wine and tea. There has been a growing interest in the investigation of the role of polyphenolic compounds in the prevention of disease conditions, such as cancer and cardiovascular diseases. Several naturally occurring plant polyphenols have been reported to inhibit the mutagenicity of chemical carcinogens by polycyclic aromatic hydrocarbons.

For example, resveratrol, a well-known polyphenol, has been reported to exert its chemoprevention activity against carcinogenesis and to provide protection against oxidative cardiovascular disorders. Most animal studies also indicate that tea exhibits chemoprevention effects against lung tumorigenesis.

There have been studies of the modulation of foreign compound-metabolic enzymes by polyphenols. It has been reported that polyphenols are capable of inducing detoxification enzymes and/or inhibition of activation enzymes. Induction of detoxification enzymes is a potential mechanism through which polyphenols carry out anticarcinogen activities.

For example, curcumin, a component of turmeric, was found to modestly induce detoxification enzyme activity in the prostate in animal models. While tea polyphenols were found to increase the activity of glutathione S-transferase.

15.3 Major Inducers of Metabolic Enzymes

Table 15.6 Enzyme modulation by polyphenols

a. Inhibition of activation enzymes	
Compound	Enzyme inhibition
Turmeric	CYP450
Curcumin	CYP450
Carnosol	CYP450
Carnosic acid	CYP450
Protocatechuic acid	CYP450
Tannic acid	CYP450
Ellagic acid	CYP450
b. Induction of detoxication enzymes	
Compound	Enzyme induction
Resveratrol	GTS, QOR, UGT
Curcumin	GTS, QOR
Turmeric	GTS
Carnosol	GTS
Carnosic acid	GTS
Protocatechuic acid	GST, UGT, NQO
Tannic acid	GST, NQO
Ellagic acid	GST, QR, UGT
Gallic acid	Phase II enzymes
Polyphenols (tea)	GST, QR

Table 15.6 lists a number of polyphenols that are reported to affect detoxification and activation enzymes. The table reveals that resveratrol is capable of inducing detoxification enzymes, while most other polyphenols induce detoxification enzymes in addition to inhibiting CYP450.

15.3.6 Organosulfur Compounds

Organosulfur compounds that have the capacity to affect activation and detoxication enzymes include diallyl sulfide, diallyl disulfide, diallyl trisulfide, and alliin. Diallyl sulfide, diallyl disulfide, and diallyl trisulfide are principal constituents of garlic oil. Among the most studied organosulfur compounds is diallyl sulfide. Diallyl sulfide was found to increase the activities of QOR and GST in the tissues of the stomach, colon, liver, lung, and urinary bladder.

Diallyl sulfide was also reported to inhibit CYP2E1 activity, but induce the activity of CYP1A1 or CYP1A2. Moreover, diallyl sulfide was shown to inhibit chemically induced carcinogenesis and cytotoxicity in animal model systems, such as the inhibition of 1,2-dimethylhydrazine–induced colon and liver cancers in rodents as well as the inhibition of arylamine N-acetyltransferase activity and gene expression in human colon cancer cell lines.

Allin, a sulfoxide, is capable of inducing UGT and GTS activities. This sulfoxide was also found to inhibit CYP2E1 activity but slightly induce CYP1A2 activity.

Table 15.7 Enzymes modulation by organosulfur compounds

a. Modulation of activation enzymes

Compound	Enzyme induction	Enzyme inhibition
Diallyl sulfide	CYP2E1	CYP1A1, CYP1A2
Diallyl disulfide	CYP2E1	
Diallyl trisulfide	–	–
Alliin	CYP2E1	CYP1A2

b. Induction of detoxification enzymes

Compound	Enzyme induction
Diallyl sulfide	GTS, QOR
Diallyl disulfide	GTS, QOR, UGT
Diallyl trisulfide	GTS, QOR
Alliin	GTS; UGT

Moreover, naturally occurring organosulfur compounds have been recognized as potential chemoprevention chemicals. For example, diallyl sulfide and diallyl disulfide were found to inhibit aflatoxin B_1–initiated carcinogenesis in rat liver.

The prospective mechanisms that are responsible for the protective effects of organosulfur compounds against chemically induced carcinogenesis are believed to be the inhibition of carcinogen activation through modulating phase I activation enzymes (such as CYP450 and monooxygenases) and/or the induction of carcinogen detoxification through inducing phase II detoxification enzymes.

Table 15.7 lists a number of organosulfur compounds that are reported to affect foreign compound–metabolizing enzymes. The table reveals that organosulfur compounds induce detoxification enzymes, but either inhibit or induce activation enzyme CYP450.

15.3.7 Terpenes and Terpenoids

Terpenes are naturally occurring hydrocarbons that are composed of various isoprene units. Among known terpenes are limonene and carotene. Limonene is a monoterpene. Lemon and citrus fruits contain a considerable amount of limonene. Foods such as carrots and cantaloupe are rich in carotenes. Carotenoids belong to a larger class of chemicals called terpenoids, which are compounds related to terpenes.

A known carotenoid is lycopene, which is present in ripe fruits, especially tomatoes. Other carotenoids include canthaxanthin, astaxanthin and zerumbone. Canthaxanthin and astaxanthin are ß-carotene-related compounds. Zerumbone is a sesquiterpene phytochemical found in subtropical edible ginger.

Table 15.8 lists a number of terpenes, terpenoids and carotenoids that are reported to modulate activation enzymes and detoxification enzymes. The table reveals that these compounds exhibit the induction of detoxification enzymes. Besides, canthaxanthin and astaxanthin also induce activation enzyme CYP450.

Table 15.8 Enzyme modulation by terpenes and terpenoids

a. Modulation of activation enzymes

Compound	Enzyme induction	Model system used for studies
Canthaxanthin	CYP450	Anima tissue
Astaxanthin	CYP450	Anima tissue

b. Induction of detoxification enzymes

Compound	Enzyme induction	Model system used for studies
ß-carotene	QOR	Animal gene
Lycopene	QOR	Animal tissue, cultured cells
Zerumbone	GST	Cultured cells
Canthaxanthin	UGT, QOR	Animal tissue
Astaxanthin	UGT; QOR	Animal tissue

Other carotenoids such as ß-carotene and astaxanthin exhibit a similar, but much smaller, effect. Moreover, the exposure of epithelial cell lines to zerumbone results in an induction of glutathione S-transferase.

15.4 Other Inducers

Foreign compound metabolic enzymes defend the body against potential harmful insults from chemicals, drugs and environments. Many xenobiotics may also induce signal transduction events, leading to various cellular, physiological and pharmacological responses. Other inducers of metabolic enzymes such as phenobarbital and grapefruit juice are described below.

15.4.1 Phenobarbital

Phenobarbital induces hepatic drug metabolic enzymes through the activation of specific nuclear receptors. Phenobarbital significantly affects the activities of hepatic CYP450, UDP-glucuronosyltransferase, glutathione S-transferase, and sulfotransferase.

15.4.2 Grapefruit

Grapefruit juice components, such as naringin and furanocoumarins, are inhibitors of activation enzyme CYP3A4. The consumption of grapefruit juice with drugs taken orally has been reported to inhibit intestinal CYP3A4 activity, which results in a decrease in the metabolism of many drugs.

Bibliography

Ahn D, Putt D, Kresty L et al (1996) The effects of dietary ellagic acid on rat hepatic and esophageal mucosal cytochromes P450 and phase II enzymes. Carcinogenesis 17:821–828

Ben-Dor A, Steiner M, Gheber L et al (2005) Carotenoids activate the antioxidant response element transcription system. Mol Cancer Ther 4:177–186

Brady JF, Ishizaki H, Fukuto JM et al (1991) Inhibition of cytochrome P-450 2E1 by diallyl sulfide and its metabolites. Chem Res Toxicol 4:642–647

Brooks JD, Paton VG, Vidanes G (2001) Potent induction of phase 2 enzymes in human prostate cells by sulforaphane. Cancer Epidemiol Biomarkers Prev 10:949–954

Chen C-H (2012) Activation and detoxification enzymes: functions and implications. Springer, New York

Chen C-H (2020) Xenobiotic metabolic enzymes: bioactivation and antioxidant defense. Springer Nature, Cham

Clark J, You M (2006) Chemoprevention of lung cancer by tea. Mol Nutr Food Res 50:144–151

Crowell PL, Gould MN (1994) Chemoprevention and therapy of cancer by d-limonene. Crit Rev Oncog 5:1–22

Davenport DM, Wargovich MJ (2005) Modulation of cytochrome P450 enzymes by organosulfur compounds from garlic. Food Chem Toxicol 43:1753–1762

Dinkova-Kostova AT, Talalay P (1999) Relation of structure of curcumin analogs to their potencies as inducers of Phase 2 detoxification enzymes. Carcinogenesis 20:911–914

Dinkova-Kostova AT, Fahey JW, Talalay P (2004) Chemical structures of inducers of nicotinamide quinone oxidoreductase 1 (NQO1). Methods Enzymol 382:423–448

Fahey JW, Talalay P (1999) Antioxidant functions of sulforaphane: a potent inducer of Phase II detoxication enzymes. Food Chem Toxicol 37:973–979

Fong AT, Swanson HI, Dashwood RH et al (1990) Mechanisms of anti-carcinogenesis by indole-3-carbinol. Studies of enzyme induction, electrophile-scavenging, and inhibition of aflatoxin B1 activation. Biochem Pharmacol 39:19–26

Fukao T, Hosono T, Misawa S et al (2004) The effects of allyl sulfides on the induction of phase II detoxification enzymes and liver injury by carbon tetrachloride. Food Chem Toxicol 42:743–749

Gradelet S, Astorg P, Leclerc J et al (1996) Effects of canthaxanthin, astaxanthin, lycopene and lutein on liver xenobiotic-metabolizing enzymes in the rat. Xenobiotica 26:49–63

Guyonnet D, Belloir C, Suschetet M et al (2001) Antimutagenic activity of organosulfur compounds from Allium is associated with phase II enzyme induction. Mutat Res 495:135–145

Hamilton SM, Teel RW (1996) Effects of isothiocyanates on cytochrome P-450 1A1 and 1A2 activity and on the mutagenicity of heterocyclic amines. Anticancer Res 16:3597–3602

Hecht SS (2000) Inhibition of carcinogenesis by isothiocyanates. Drug Metab Rev 32:395–411

Kang JJ, Wang HW, Liu TY et al (1997) Modulation of cytochrome P-450-dependent monooxygenases, glutathione and glutathione S-transferase in rat liver by geniposide from Gardenia jasminoides. Food Chem Toxicol 35:957–965

Krajka-Kuźniak V, Szaefer H, Baer-Dubowska W (2004) Modulation of 3-methylcholanthrene-induced rat hepatic and renal cytochrome P450 and phase II enzymes by plant phenols: protocatechuic and tannic acids. Toxicol Lett 152:117–126

Kwak MK, Egner PA, Dolan PM et al (2001) Role of phase 2 enzyme induction in chemoprotection by dithiolethiones. Mutat Res 480-481:305–315

Lançon A, Hanet N, Jannin B et al (2007) Resveratrol in human hepatoma Hep G2 cells: metabolism and inducibility of detoxifying enzymes. Drug Metab Dispos 35:699–703

Li Y, Cao Z, Zhu H (2006) Upregulation of endogenous antioxidants and phase 2 enzymes by the red wine polyphenol, resveratrol in cultured aortic smooth muscle cells leads to cytoprotection against oxidative and electrophilic stress. Pharmacol Res 53:6–15

Li Y, Mezei O, Shay NF (2007) Human and murine hepatic sterol-12-alpha-hydroxylase and other xenobiotic metabolism mRNA are upregulated by soy isoflavones. J Nutr 137:1705–1712

Maheo K, Morel F, Langouet S et al (1997) Inhibition of cytochromes P-450 and induction of glutathione S-transferases by sulforaphane in primary human and rat hepatocytes. Cancer Res 57:3649–3652

Moon YJ, Wang X, Morris ME (2006) Dietary flavonoids: effects on xenobiotic and carcinogen metabolism. Toxicol In Vitro 20:187–210

Munday R, Munday CM (2001) Relative activities of organosulfur compounds derived from onions and garlic in increasing tissue activities of quinone reductase and glutathione transferase in rat tissues. Nutr Cancer 40:205–210

Munday R, Zhang Y, Paonessa JD et al (2010) Synthesis, biological evaluation, and structure-activity relationships of dithiolethiones as inducers of cytoprotective phase 2 enzymes. J Med Chem 53:4761–4767

Nakajima M, Yoshida R, Shimada N et al (2001) Inhibition and inactivation of human cytochrome P450 isoforms by phenethyl isothiocyanate. Drug Metab Dispos 29:1110–1113

Nakamura Y, Yoshida C, Murakami A et al (2004) A tropical ginger sesquiterpene, activates phase II drug metabolizing enzymes. FEBS Lett 572:245–250

Ow YY, Stupans I (2003) Gallic acid and gallic acid derivatives: effects on drug metabolizing enzymes. Curr Drug Metab 4:241–248

Pugazhenthi S, Akhov L, Selvaraj G et al (2007) Regulation of heme oxygenase-1expression by demethoxy curcuminoids through Nrf2 by a PI3-kinase/Akt-mediated pathway in mouse beta-cells. Am J Physiol Endocrinol Metab 293:E645–E655

Ramos-Gomez M, Kwak MK et al (2001) Sensitivity to carcinogenesis is increased and chemoprotective efficacy of enzyme inducers is lost in nrf2 transcription factor-deficient mice. Proc Natl Acad Sci USA 98:3410–3415

Reicks MM, Crankshaw DL (1996) Modulation of rat hepatic cytochrome P-450 activity by garlic organosulfur compounds. Nutr Cancer 25:241–248

Roebuck BD, Curphey TJ, Li Y et al (2003) Evaluation of the cancer chemopreventive potency of dithiolethione analogs of oltipraz. Carcinogenesis 24:1919–1928

Rogan EG (2006) The natural chemopreventive compound indole-3-carbinol: state of the science. In Vivo 20:221–228

Rushmore TH, Kong A-NT (2002) Pharmacogenomics, regulation and signaling pathways of phase I and II drug metabolizing enzymes. Curr Drug Metab 3(5):481–490

Seo K, Jung S, Park M et al (2001) Effects of leucocyanidines on activities of metabolizing enzymes and antioxidant enzymes. Biol Pharm Bull 24:592–593

Shih PH, Yeh CT, Yen GC (2007) Anthocyanins induce the activation of phase II enzymes through the antioxidant response element pathway against oxidative stress-induced apoptosis. J Agric Food Chem 55:9427–9435

Song LL, Kosmeder JW 2nd, Lee SK et al (1999) Cancer chemopreventive activity mediated by 4′-bromoflavone, a potent inducer of phase II detoxification enzymes. Cancer Res 59:578–585

Steele VE, Kelloff GJ, Balentine D et al (2000) Comparative chemopreventive mechanisms of green tea, black tea and selected polyphenol extracts measured by in vitro bioassays. Carcinogenesis 21:63–67

Tsai CW, Chen HW, Yang JJ et al (2007) Diallyl disulfide and diallyl trisulfide up-regulate the expression of the pi class of glutathione S-transferase via an AP-1-dependent pathway. J Agric Food Chem 55:1019–1026

von Weymarn LB, Chun JA, Knudsen GA et al (2007) Effects of eleven isothiocyanates on P450 2A6- and 2A13-catalyzed coumarin 7-hydroxylation. Chem Res Toxicol 20:1252–1259

Wallig MA, Kingston S, Staack R et al (1998) Induction of rat pancreatic glutathione S-transferase and quinone reductase activities by a mixture of glucosinolate breakdown derivatives found in Brussels sprouts. Food Chem Toxicol 36:365–373

Wang W, Liu LQ, Higuchi CM et al (1998) Induction of NADPH:quinone reductase by dietary phytoestrogens in colonic Colo205 cells. Biochem Pharmacol 56:189–195

Xu M (1999) Dashwood RH. Chemoprevention studies of heterocyclic amine-induced colon carcinogenesis. Cancer Lett 143:179–183

Yang CS, Chhabra SK, Hong JY et al (2001) Mechanisms of inhibition of chemical toxicity and carcinogenesis by diallyl sulfide (DAS) and related compounds from garlic. J Nutr 131:1041S–1045S

Yannai S, Day AJ, Williamson G et al (1998) Characterization of flavonoids as monofunctional or bifunctional inducers of quinone reductase in murine hepatoma cell lines. Toxicology 36:623–630

Zancanella V, Giantin M, Lopparelli RM et al (2012) Constitutive expression and phenobarbital modulation of drug metabolizing enzymes and related nuclear receptors in cattle liver and extra-hepatic tissues. Xenobiotica 42(11):1096–1109

Zhang W, Go ML (2007) Quinone reductase induction activity of methoxylated analogues of resveratrol. Eur J Med Chem 42:841–850

Diversified Classes of Enzyme Modulators 16

The structure of an enzyme is required to stabilize the conformation of the active site for achieving enzymatic function. The active site of an enzyme is usually larger than the substrate.

Enzymatic catalysis is mediated by functional groups of amino acid side chains, including histidine, serine, cysteine, lysine, glutamate, and aspartate. In a metabolic reaction involving a chemical change in the parent compound, enzymatic catalysis could not be brought about only by the functional group present in the amino acid side chain alone. The enzyme also needs to act in cooperation with a small substrate molecule.

An efficient condition for an enzyme catalytic action is to have the activity of the enzyme proportional to its need at any particular time. In the circumstance of the accumulation of foreign compounds, the cells must be able to detoxify them by appropriate metabolic enzymes in order to avoid their potential toxic effects.

An important feature of foreign compound- metabolic enzymes is their ability to be induced by a variety of chemical compounds referred to as modulators. A modulator is a small molecule that binds to the enzyme, either covalently or noncovalently, thereby changing the structure or conformation of the enzyme in a manner that the enzyme activity is either increased or decreased.

Prior to exerting its action on the enzyme present in the cells, a modulator must be able to penetrate across cell membranes. Extensive research indicates that substantial protection against xenobiotic-mediated toxic effects or chemical carcinogenesis can be achieved by induction of metabolic enzymes for metabolism of foreign compounds, including foods, drugs, and environmental and lifestyle chemicals.

16.1 Substrate–Enzyme Interactions

Foreign compounds or their metabolites are the substrates for xenobiotic metabolic enzyme–catalyzed reactions. These substrates are either electrophiles or nucleophiles. Electrophiles are electron-deficient substances that interact with electron-rich species, which are considered as Lewis acids. Electrophilic metabolites contain either positively charged ions such as H^+ and NO^+ or molecules that carry partial positive charge or polarized neutral molecules, such as alkyl halides, acyl halides, and carbonyl compounds.

Nucleophilic compounds are electron-rich substances that interact with electron-deficient species. Having an excess in electrons, nucleophiles are considered as Lewis bases. Electron-rich nucleophilic metabolites include many compounds that contain O, N, or S atoms or functional groups such as -OH, -NH$_2$, -SH, and -COOH.

Metabolic intermediates generated through metabolic activation catalyzed by phase I activation enzymes are substrates for phase II enzymes. Such substrates often share a common feature that contains electrophilic atoms. Most of reactive active intermediates are electrophiles. Electrophilic metabolites prefer to interact with nucleophilic groups of enzymes.

16.1.1 Electrophilic and Nucleophilic Groups

Table 16.1 presents a number of electrophilic and nucleophilic groups, particularly serine and tyrosine hydroxyl, aspartate or glutamate carboxylase, histidine imidazole, and cysteine sulfhydryl groups. Nucleophilic metabolites prefer to interact with electrophilic groups in the amino acid side chains of xenobiotic metabolic enzymes. While, a number of potentially electrophilic groups in metabolic enzymes are such as -NH$_3^+$ and metal ions Mg^{2+}, Mn^{2+} or Fe^{3+}.

16.1.2 Conjugation of Metabolite

Glucoronidation is the primary metabolic reaction for many compounds containing nucleophilic functional groups, such as -OH, -COOH, -SH, and -NH$_2$. For example, in uridine-diphosphate glucuronosyltransferase–catalyzed reactions, the site of glucuronidation is generally an electron-rich nucleophilic O, N, or S atom. While sulfonation conjugation reactions involve the transfer of a sulfonate group (-SO$_3^-$)

Table 16.1 Electrophilic and nucleophilic atoms or groups in enzymes and substrates

Substrates		Enzymes	
Electrophiles	Nucleophiles	Electrophiles	Nucleophilic group
H^+	O atom	-NH$_3^+$	Serine
NO^+	N atom	Mg^{2+}	Tyrosine hydroxyl
Alkyl halides	S atom	Mn^{2+}	Aspartate carboxylase
Acyl halides	-OH group	Fe^{3+}	Glutamate carboxylase
Carbonyl compounds	-NH$_2$ group		Histidine imidazole
	-SH group		Cysteine sulfhydryl
	-COOH group		

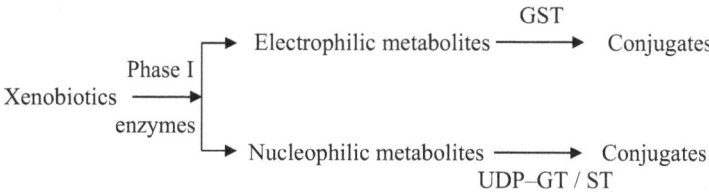

Fig. 16.1 Conjugations of metabolites by enzyme-catalyzed reactions

from the cofactor (3-phosphoadenosine 5- phosphosulfate) as the donor to a nucleophilic group of a metabolite intermediate (as the acceptor).

Figure 16.1 illustrates conjugation of metabolites by enzyme-catalyzed reactions.

Electrophilic metabolites generated by phase I activation enzymes are detoxified by phase II detoxification enzymes, such as glutathione S-transferases. While nucleophilic metabolites, such as phenols, are often detoxified by uridine-diphosphate glucuronosyltransferases and sulfotransferases.

In nucleophilic catalysis, the roles of catalyst and substrate are the reverse of those for electrophilic catalysis. Detoxification of electrophiles is an important event, since most reactive metabolites consist of electrophiles. However, the detoxification of nucleophiles is as important, since many nucleophiles can be converted to electrophiles.

16.2 Interaction of Modulator with Metabolic Enzyme

Foreign compound metabolisms require the substrate–enzyme interactions, which involve the reactive group of the metabolite and the function group in the amino side chains of the enzyme. Through either covalent or noncovalent interaction, a modulator molecule is able to affect the activity of foreign compound metabolic enzyme.

Investigations of substrate–enzyme interactions in the presence of a modulator are essential in the elucidation of the mechanism underlying the modulator effects on the enzyme activity. Such an interference may cause the activity of the enzyme to be increased or decreased, depending on whether the interference results in more or less favorable substrate–enzyme interactions.

For example, benzo[a]pyrene (BaP) is a human carcinogen requiring metabolic activation prior to reaction with DNA. BaP induces CYP1A enzyme activity, resulting an increase in the formation of BaP-DNA adducts. Besides BaP, CYP1A1 is strongly induced by many different chemical agents, including 2,3,7,8-tetrachlorodibenzo-*p*-dioxin, which binds to the aryl hydrocarbon receptor.

16.2.1 Enzyme–Substrate Interaction

A diversity of small molecules of naturally occurring or synthetic origins has been found to be effective inducers of phase II detoxification enzymes. They have the potential to protect organisms against foreign compound–mediated toxic effects. Induction of phase II detoxification enzyme is considered as a major strategy for reducing the susceptibility of living cells to toxic and chemopreventive effects of foreign compound metabolites.

For example, the potential of isothiocyanates to protect against the carcinogenicity of reactive chemical intermediates involves the impair of CYP450 activity as well as the induction of detoxification enzyme systems, including quinone reductase, glutathione S-transferase, epoxide hydrolase, and UDP-glucuronosyl transferase.

Evaluation of the chemical structures of such small modulation molecules reveals that enzyme modulators largely belong to a variety of chemical classes with few common properties, except their ability to modify sulfhydryl group of cysteine residue of the enzymes.

16.2.2 Enzyme Conformation

A modulator molecule may also be capable of affecting the activity of metabolic enzyme by altering its conformation. The conformation of enzyme is crucial for the enzymatic function. Understanding the potential effect of a modulator on the enzyme conformation is essential to the elucidation of the mechanisms underlying the induction of foreign compound-metabolic enzymes.

Fluorescence spectroscopy is useful for investigating the conformation of enzyme upon the binding of a small molecule. Stop-flow kinetics is valuable in elucidating the mechanism of small molecule–enzyme interaction. Future applications of fluorescence spectroscopy and stop-flow kinetics are needed to study the effects of modulators on substrate–enzyme interactions and the conformation of enzymes.

16.3 Michael Acceptor Functionalities

Extensive research has revealed that besides their ability to modify sulfhydryl group of a cysteine residue, many modulators of phase II detoxification enzymes contain Michael acceptor functionalities, such as olefins or acetylenes conjugated to electron withdrawing groups. Olefins are unsaturated hydrocarbons that contain one or more pairs of carbon atoms linked by a double bond.

Acetylenes are also unsaturated hydrocarbons, but consist of two carbon atoms linked by a triple bond. Carbon–carbon double bond in olefins or carbon–carbon triple bond in acetylene serves as a source of electrons. An electron-withdrawing group attached to the carbon–carbon double or the carbon–carbon triple bond can destabilize the transition state of Michael acceptor.

The potency of these enzyme modulators was found to parallel their reactivity as Michael acceptors.

16.4 Enzyme Modulators with Michael Acceptor Characteristics

C=O, -COOH, -COOR, and -CN groups that attach to the carbon–carbon double bond are powerful electron withdrawing groups. For example, the carbon–carbon double bond of an α,β -unsaturated ketone, acid, ester, or nitrile is susceptible to nucleophilic attack. Michael reaction acceptor is the addition of nucleophile (electron-rich Michael donor) to an unsaturated carbonyl compound (electron-deficient Michael acceptor).

Functional groups of Michael reaction receptors include α,β-unsaturated double bond attached to aldehydes, ketone, quinone, thioketone, sulfone, ester, nitrile, or nitro groups. A list of enzyme modulators that contain Michael acceptor functionalities is presented in Table 16.2.

Table 16.2 A list of enzyme modulators containing Michael acceptors

Classes of modulators	Compounds
Sulforaphane and isothiocyanate	Sulforaphane
	Phenylethyl isothiocyanate
1,2-dithiole-3-thione and derivatives	1,2-dithioe-3-thione (D3T)
	4-methyl-5-pyrazinyl-D3T (OPZ)
Indole-3-carbinol	Indole-3-carbinol
Flavonoids	Catechin, epicatechin
	Epigallocatechin, leucocyanidines Myricetin, quercetin
	Fisetin, isoliquiritigenin
	Diosmin, hesperidin
Isoflavones	Genistein, daidzein
Phenols and polyphenols	Resveratrol, curcumin
	Gallic acid, rosmarinic acid
	Carnosic acid, tannin
	Ellagic acid, protocatechuic acid
Organosulfur	Diallyl sulfide, diallyl disulfide
	Diallyl trisulfide, Alliin
Terpenes and terpenoids	Beta-carotene, lycopene
	Canthaxanthin, astaxanthin
	Zerumbone, limonene
Quinoline	Ethoxyquinn
Others	Nivalenol

16.5 Diversities of Enzyme Inducers

Some phase I enzyme metabolisms are accountable for the activation of foreign compounds to chemically reactive intermediates. Substantial evidences indicate that significant protection against chemical carcinogenesis and inflammatory conditions can be achieved by induction of enzymes responsible for the metabolism of such xenobiotic reactive metabolites.

Due to selective phase II enzyme inducers that offer a potential for achieving such protections, it has been proposed that induction of phase II activation enzymes is a sufficient condition for chemoprotection. In regard to the protective role of inducers, structural features of enzyme inducers have a significant impact. Enzyme modulators that contain Michael Acceptors are listed in Table 16.2.

16.5.1 Ortho-Hydroxyl Group on Aromatic Ring

Ortho-hydroxyl groups have a significant impact on the protective role of phase II enzyme inducers. For example, the introduction of ortho-hydroxyl groups on the aromatic rings of phenylpropenoids was found to dramatically enhance their potencies, not only as inducers of quinone reductase but also as quenchers of superoxide. The presence of an ortho-hydroxyl substituent on the aromatic ring was also reported to profoundly increase the induction potency of benzylidene-alkanones and -cycloalkanones.

A list of enzyme inducers that contain ortho-hydroxyl groups on the aromatic rings is shown in Table 16.3. Besides Michael acceptor functionalities, the potencies of quinone reductase induction appear to be correlated with their ability to quench superoxide radicals. The involvement of both Michael reaction reactivity and radical quenching mechanisms suggests that these inducers are bifunctional antioxidants.

16.5.2 Chemical Structures of Enzyme Inducers

There are diversified classes of enzyme inducers. Many of phase II enzyme inducers contain Michael acceptor functionalities. To help in understanding the functional

Table 16.3 Enzyme modulators containing ortho-hydroxyl group on the aromatic ring

Classes of compound	Enzyme modulator
Flavonoids	Catechin, epicatechin, myricetin,
	Epigallocatechin, leucocyanidines
	Quercetin, fisetin, diosmin, hesperidin
Phenols and polyphenols	Curcumin, gallic acid, tannin,
	Rosmarinic acid, carnosic acid
	Ellagic acid, protocatechuic acid

16.5 Diversities of Enzyme Inducers

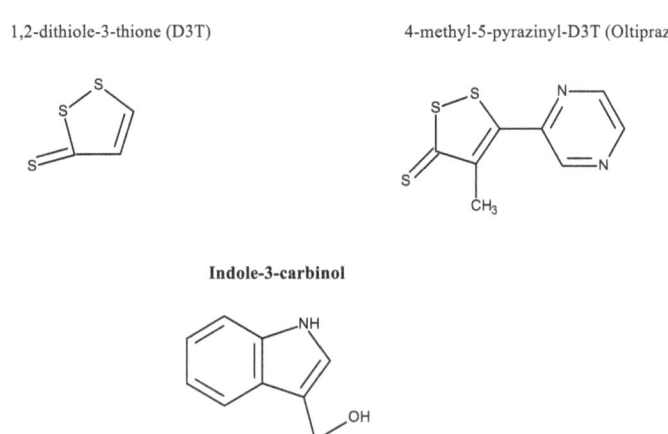

Fig. 16.2 Chemical structures of phase II enzyme inducers: Sulforaphane, phenylethyl isothiocyanate, 1,2-dithiole-3-thione, and oltipraz

characteristics of enzyme inducers, the chemical structures of a variety of enzyme inducers are shown in Figs. 16.2, 16.3, 16.4, 16.5, 16.6, 16.7, 16.8 and 16.9.

In addition to Michael acceptor functionalities, some inducers of phase II enzymes, such as flavonoid and curcuminoid analogues, contain phenolic hydroxyl groups. These phenol hydroxyl groups can scavenge reactive oxygen and nitrogen species. Moreover, flavonoids and curcuminoids not only play an indirect protective role by inducing phase II enzymes but also a direct protective role by scavenging hazardous oxidants. Furthermore, enzyme inducers that contain phenolic hydroxyl groups in addition to Michael acceptor centers play not only indirect, but also direct protective roles as bifunctional antioxidants.

Flavonoids (1)

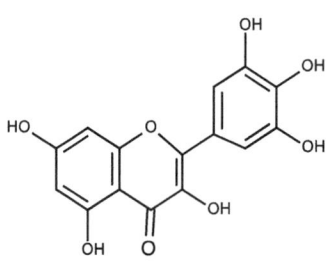

Fig. 16.3 Chemical structures of phase II enzyme inducers: Catechin, epicatechin, epigallocatechin, leucocyanidin, myricetin, and quercetin

(a) Sulforaphane, Phenylethyl Isothiocyanate, 1,2-Dithiole-3-Thione, and Oltipraz

The chemical structures of sulforaphane, phenylethyl isothiocyanate, 1,2-dithiole-3-thione, and oltipraz are shown in Fig. 16.2.

16.5 Diversities of Enzyme Inducers

Flavonoids (2)

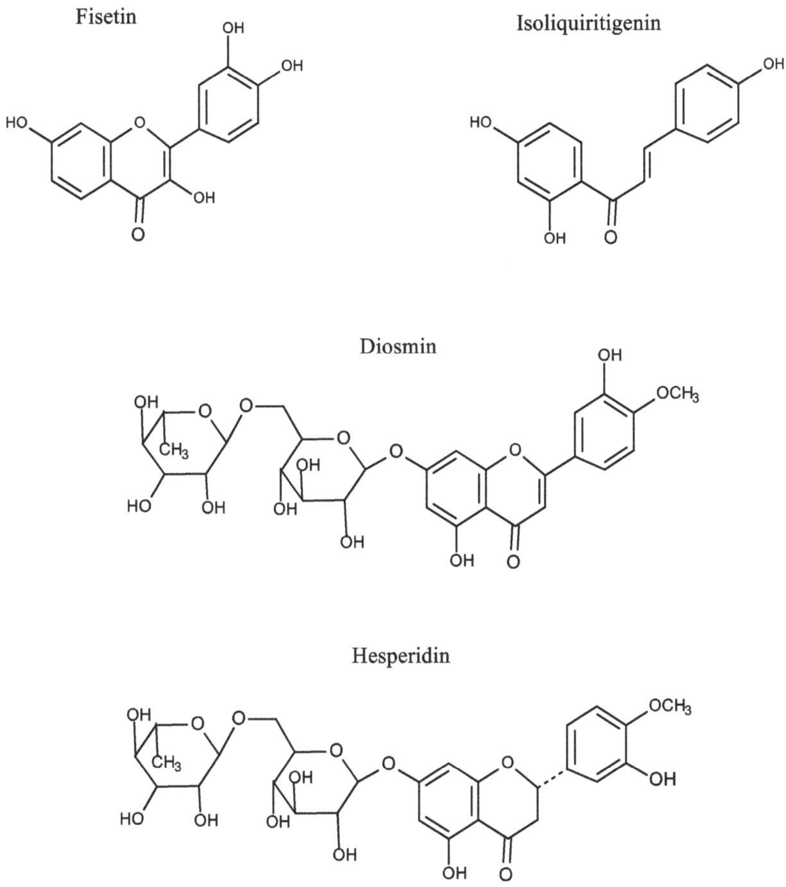

Fig. 16.4 Chemical structures of phase II enzyme inducers: Fisetin, isoliquiritigenin, diosmin, and hesperidin

(b) Catechin, Epicatechin, Epigallocatechin, Leucocyanidin, Myricetin, and Quercetin

The Chemical structures of catechin, epicatechin, epigallocatechin, leucocyanidin, myricetin, and quercetin are shown in Fig. 16.3.

(c) Fisetin, Isoliquiritigenin, Diosmin, and Hesperidin

The chemical structures of fisetin, isoliquiritigenin, diosmin, and hesperidin are shown in Fig. 16.4.

Isoflavones

Genistein

Daidzein

Phenols and polyphenols (1)

Resveratrol

Curcumin

Gallic acid

Rosmarinic acid

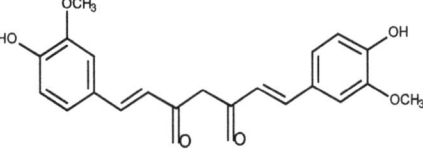

Fig. 16.5 Chemical structures of phase II enzyme inducers: Genistein, daidzein, resveratrol, curcumin, gallic acid, and rosmarinic acid

(d) Genistein, Daidzein, Resveratrol, Curcumin, Gallic Acid, and Rosmarinic Acid

The chemical structures of genistein, daidzein, resveratrol, curcumin, gallic acid, and rosmarinic acid are shown in Fig. 16.5.

(e) Phenols, Polyphenols, Ellagic Acid, and Protocatechuic Acid

The chemical structures of phenols, polyphenols, ellagic acid, and protocatechuic acid are shown in Fig. 16.6.

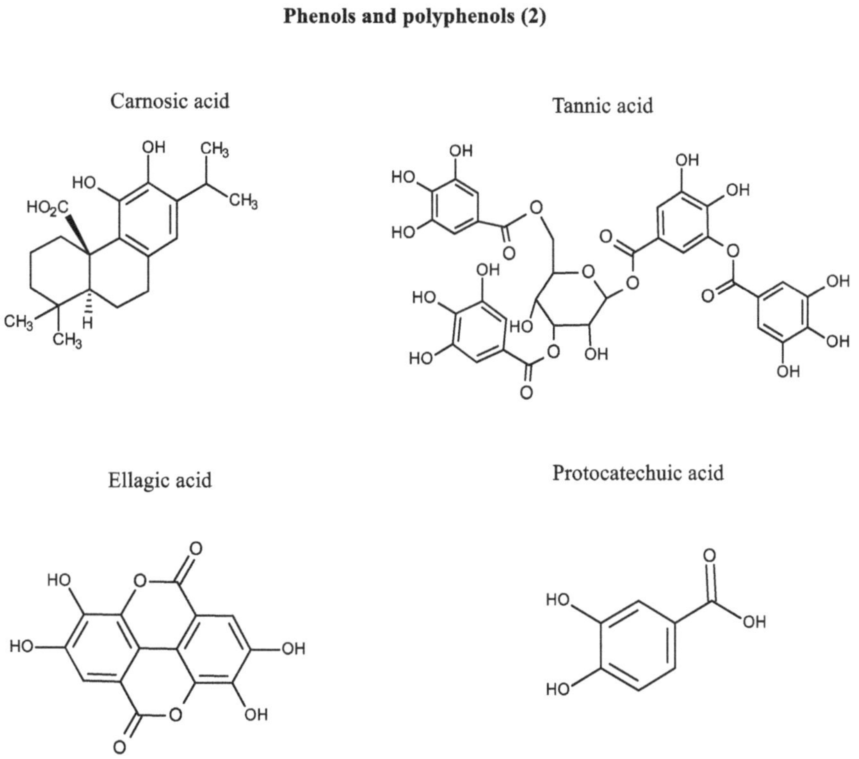

16.5 Diversities of Enzyme Inducers

Phenols and polyphenols (2)

Carnosic acid

Tannic acid

Ellagic acid

Protocatechuic acid

Fig. 16.6 Chemical structures of phase II enzyme inducers: Phenols, polyphenols, ellagic Acid, and protocatechuic acid

Organosulfur compounds

Diallyl sulfide

Diallyl disulfide

Diallyl trisulfide

Alliin

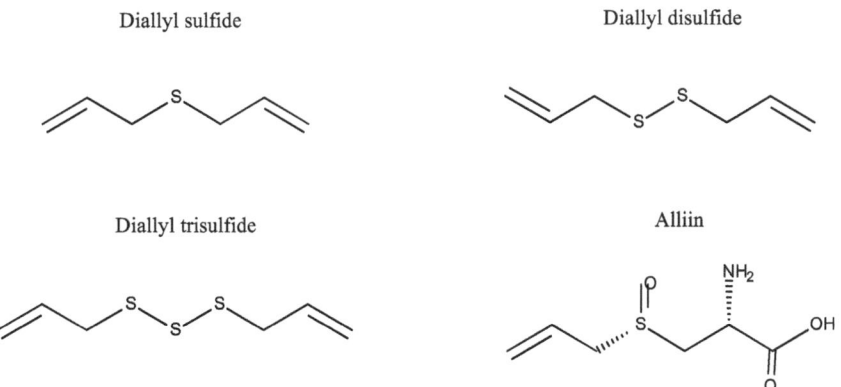

Fig. 16.7 Chemical structures of phase II enzyme inducers: Dially sulfide, dially disulfide, dially trisulfide, and alliin

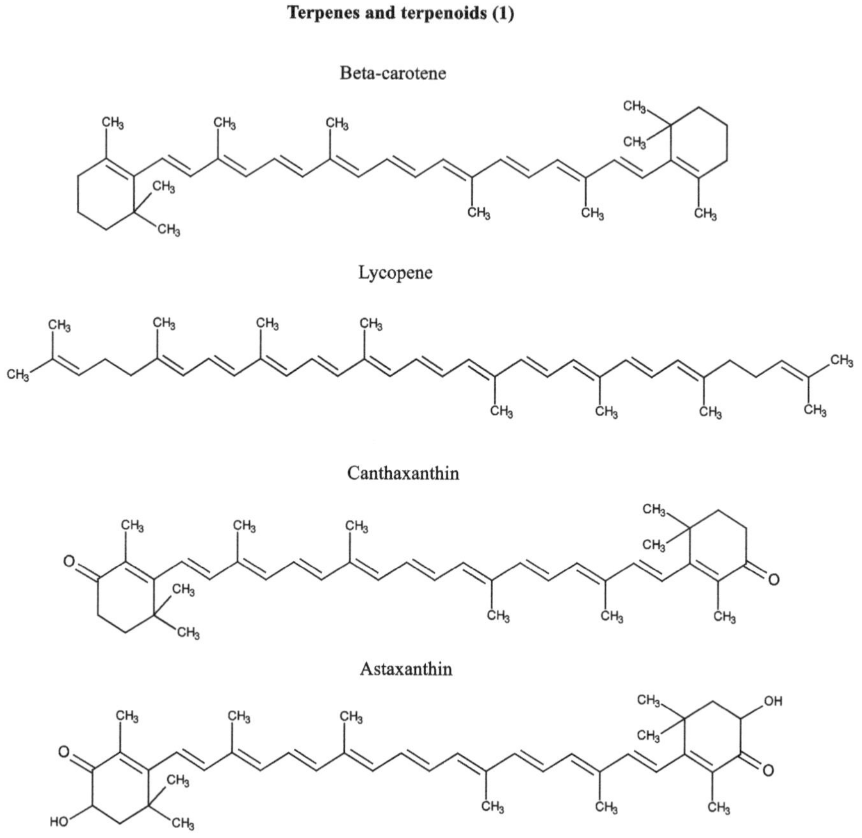

Terpenes and terpenoids (1)

Beta-carotene

Lycopene

Canthaxanthin

Astaxanthin

Fig. 16.8 Chemical structures of phase II enzyme inducers: Beta-carotene, lycopene, canthaxanthin, and astaxanthi

(f) Dially Sulfide, Dially Disulfide, Dially Trisulfide, and Alliin

The chemical structures of dially sulfide, dially disulfide, dially trisulfide, and alliin are shown in Fig. 16.7.

(g) Beta-carotene, Lycopene, Canthaxanthin, and Astaxanthin

The chemical structures of beta-carotene, lycopene, canthaxanthin, and astaxanthin are shown in Fig. 16.8.

(h) Zerumbone, Limonene, and Ethoxyquin

The chemical structures of zerumbone, limonene, and ethoxyquin are presented in Fig. 16.9.

Terpenes and terpenoids (2)

Zerumbone

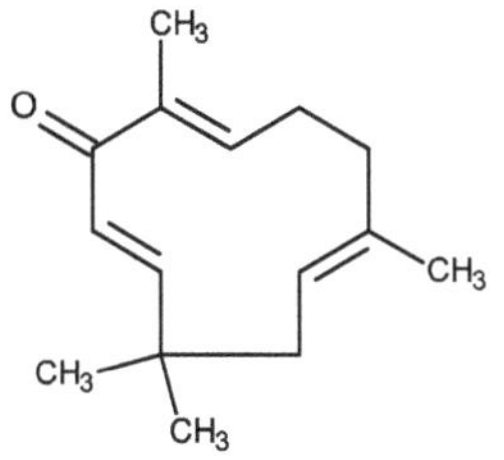

Limonene

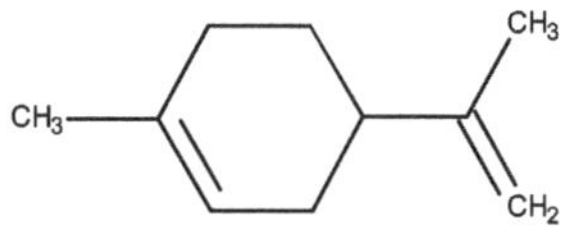

Quinoline

Ethoxyquin

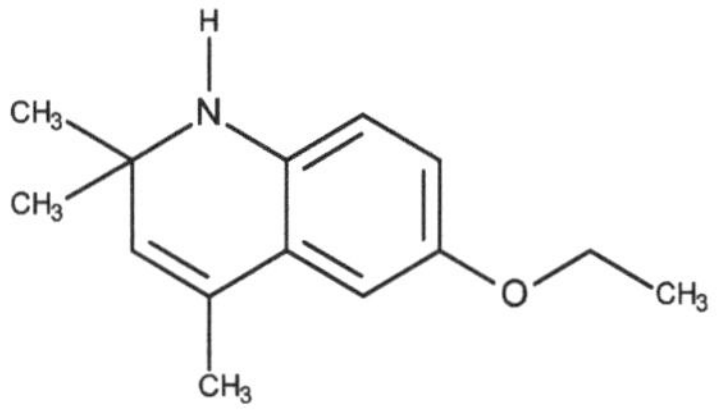

Fig. 16.9 Chemical structures of phase II enzyme inducers: Zerumbone, limonene, and ethoxyquin

Bibliography

Bock KW, Lilienblum W, Fischer G et al (1987) The role of conjugation reactions in detoxication. Arch Toxicol 60:22–29

Bolton JL, Trush MA, Penning TM et al (2000) Role of quinones in toxicology. Chem Res Toxicol 13:135–160

Chen C-H (n.d.-a) Activation and Detoxification Enzymes: Functions and Implications. Springer Sciences, New York

Chen C-H (n.d.-b) Xenobiotic Metabolisms: Bioactivation and Antioxidant Defense. Springer Nature, Switzerland

Ciaccio PJ, Jaiswal AK, Tew KD (1994) Regulation of human dihydrodiol dehydrogenase by Michael acceptor xenobiotics. J Biol Chem. 269(15558–15):562

Cuendet M, Oteham CP, Moon RC et al (2006) Quinone reductase induction as a biomarker for cancer chemoprevention. J Nat Prod. 69:460–463

Dinkova-Kostova AT, Abeygunawardana C, Talalay P (1998) Chemoprotective properties of phenylpropenoids, bis(benzylidene)cycloalkanones, and related Michael reaction acceptors: correlation of potencies as phase 2 enzyme inducers and radical scavengers. J. Med Chem. 41:5287–5296

Dinkova-Kostova AT, Cheah J, Samouilov A et al (2007) Phenolic Michael reaction acceptors: combined direct and indirect antioxidant defenses against electrophiles and oxidants. Med Chem 3:261–268

Dinkova-Kostova AT, Holtzclaw WD, Cole RN et al (2002) Direct evidence that sulfhydryl groups of Keap1 are the sensors regulating induction of phase 2 enzymes that protect against carcinogens and oxidants. Proc Nat Acad Sci 99(11908–11):913

Dinkova-Kostova AT, Massiah MA, Bozak RE et al (2001) Potency of Michael reaction acceptors as inducers of enzymes that protect against carcinogenesis depends on their reactivity with sulfhydryl groups. Proc Natl Acad Sci 98:3404–3409

Hodek P, Koblihova J, Kizek R et al (2013) The relationship between DNA adduct formation by benzo[a]pyrene and expression of its activation enzyme cytochrome P450 1A1 in rat. Environ Toxicol Pharmacol 36(3):989–996

Nordlund P, Reichard P (2006) Ribonucleotide reductases catalyze the substitution of the 2 OH-group of a ribonucleotide with a hydrogen. Conformational transitions induced by nucleotide binding determine the regulation of substrate specificity. Annu Rev. Biochem 75:681–706

Prochaska HJ, Talalay P (1988) Regulatory mechanisms of monofunctional and bifunctional anticarcinogenic enzyme inducers in murine liver. Cancer Res. 48:4776–4782

Razis AFA, Konsue N, Ioannides C (2018) Isothiocyanates and xenobiotic detoxification. Mol Nutr Food Res 62(18):e1700916

Rinaldi R, Eliasson E, Swedmark S et al (2002) Reactive intermediates and the dynamics of glutathione transferases. Drug Metab Dispos. 30:1053–1058

Schultz TW, Yarbrough JW, Hunter RS et al (2007) Verification of the structural alerts for Michael acceptors. Chem Res Toxicol. 20:1359–1363

Shiizaki K, Kawanishi M, Yagi T (2017) Modulation of benzo[a]pyrene-DNA adduct formation by CYP1 inducer and inhibitor. Genes Environ 39:14

Talalay P (1989) Mechanisms of induction of enzymes that protect against chemical carcinogenesis. Adv Enzyme Regul. 28:237–250

Talalay P, De Long MJ, Prochaska HJ (1988) Identification of a common chemical signal regulating the induction of enzymes that protect against chemical carcinogenesis. Proc Natl Acad Sci. 85:8261–8265

Zhang F, Thottananiyil M, Martin DL, Chen CH (1999) Conformational alteration in serum albumin as a carrier for pyridoxal phosphate: a distinction from pyridoxal phosphate-dependent glutamate decarboxylase. Arch Biochem Biophys. 364:195–202

Metabolite-Associated Disease Conditions 17

As a result of an imbalance between activation and detoxification metabolisms, reactive chemical intermediates and oxidative stress occur, leading to a loss of control over redox signaling processes and causing potential biomolecular damages. When protective defenses are overwhelmed by excess toxicant insult, reactive chemical intermediates can lead to deregulation of cell signaling pathways and dysfunction of biomolecules.

Oxidative stress is involved in the causes of many disease conditions, varying from diabetes, neurodegeneration to cancer. Research aims at elucidating that the molecular mechanism of foreign compound-induced toxicities is essential for preventing such organ injuries. Extensive research is needed to prevent or devise treatments for injuries caused by reactive chemical intermediates produced by foreign compound metabolism.

Most drugs are hydrophobic substances that are able to enter through lipid bilayers into cells, where drugs interact with their target receptors. Xenobiotic metabolic enzymes convert drugs into compounds with hydrophilic derivatives, which are more easily eliminated through excretion into the aqueous compartments of the tissues. During such conversion, some drugs undergo metabolic activation to form reactive metabolites.

To address the issue of oxidative stress derived from metabolic reactive intermediates, nuclear factor erythroid 2-related factor 2 (NRF2), a transcription factor, plays an important oxidative stress regulator. Nrf2 acts as an essential role in the expression of cytoprotective genes involved in the antioxidant and anti-inflammatory responses. The modulation of Nrf2 is a promising approach in the prevention and treatment of reactive metabolite-associated toxicities.

17.1 Metabolite-Associated Hepatotoxicity

In order to avoid accumulation in the body to cause potential toxic effects, foreign compounds that enter into the body are eliminated through metabolic mechanisms and then excreted in the urine or bile. The liver plays a prominent role in the metabolism of xenobiotics, where reactive chemical intermediates are converted into water-soluble metabolites so that xenobiotics can be eliminated from the body through urine. Such conversion is carried out by metabolisms catalyzed by activation and detoxification enzymes.

Chemical-induced liver injuries are caused by reactive intermediate species, such as free radicals or electrophilic compounds formed during metabolic conversion of foreign compounds. Besides metabolic reactive intermediates, genetic, environmental, and lifestyle factors also determine the susceptibility of individuals to such metabolite-induced injuries.

Many reactive intermediate species can generate and lead to oxidative stress, causing cellular damages. For examples, hepatotoxic and carcinogenic species can be generated by metabolic hepatic enzymes, such as CYP2B6 and CYP3A4. Their potential mechanisms underlying hepatotoxicity are related to intracellular glutathione depletion or quinone formation.

Liver is the main site for biotransformation of xenobiotics. Such a critical role makes liver particularly susceptible to injury by exposing to foreign compounds. The pathogenesis of most chemical-induced hepatotoxicity is initiated by electrophilic compound or free radical-induced oxidative stress, which can potentially alter the structure and function of cellular macromolecules.

Besides acting as solute carrier transporters that mediate influx or efflux of foreign compounds, the liver also plays an important role in removing foreign compounds from blood, after they are absorbed in the gastrointestinal tract. Hepatic uptakes of anions, cations, and salts are carried out by either facilitated diffusion or active transport. In active transport, ATP-binding cassette transport transporters in hepatocytes mediate efflux of foreign compounds such as drugs, where the driving force is ATP hydrolysis.

17.1.1 Alcohol and Aldehyde

Chronic alcohol consumption is a well-known risk factor for liver disease. The oxidative damage induced by ethanol significantly contribute to the pathogenesis of alcohol hepatotoxicity.

Alcohol affects the liver through metabolic disturbances associated with its oxidation. Redox changes produced by the hepatic alcohol dehydrogenase pathway affect lipid, carbohydrate, and protein metabolisms.

Ethanol is oxidized in liver microsomes by the ethanol-inducible cytochrome CYP2E1, resulting in hepatic damage. Aldehyde dehydrogenase is the major enzyme that metabolizes acetaldehyde produced from alcohol metabolism. Aldehyde dehydrogenase gene leads to decrease in oxidative stress and inflammation during ethanol-induced liver damage. People who carry an inactive aldehyde dehydrogenase gene can exhibit acetaldehyde accumulation and greater hepatic inflammation after alcohol consumption.

The toxicity and carcinogenicity of formaldehyde (HCHO) has been attributed to its ability to form adducts with proteins and DNA. Moreover, when isolated hepatocytes are incubated with formaldehyde, a marked decrease in mitochondrial membrane potential and inhibition of mitochondrial respiration can occur. Such marked decrease in mitochondrial membrane potential and inhibition of mitochondrial respiration was found to be accompanied by reactive oxygen species formation.

Methanol is metabolized to toxic compounds by enzymatic pathways. Understanding methanol-induced cytotoxicity could provide insight into the molecular basis for acute methanol-induced hepatotoxicity. Hepatocyte protein carbonylation induces formaldehyde formation. Increased HCHO concentration levels are correlated with increased HCHO-induced protein carbonylation in hepatocytes.

Furthermore, cytotoxicity studies were conducted in isolated hepatocytes, which were exposed to HCHO and were treated with inhibitors of HCHO metabolizing enzymes. Inhibition of glutathione (GSH)-dependent HCHO dehydrogenase activity by prior depletion of GSH was found to markedly increase the susceptibility of hepatocyte to HCHO.

17.1.2 Aflatoxin-Induced Hepatic Carcinogenesis

Hepatocellular carcinoma is influenced by aflatoxin B1 metabolite. Reactive oxygen species can play a leading role in initiation and promotion of hepatic carcinogenesis. When a drug is targeted for phase I oxidation through the CYP450s, its rate of elimination depends on the initial phase I oxidation reaction.

Moreover, if the catalytic rates of phase II reactions are significantly faster than the rates of the CYP450s, phase II reactions are generally considered to assume the efficient elimination and detoxification of most drugs. For example, phase II detoxification reactions of glucuronidation and sulfation result in the formation of conjugates with a significant increase in water solubility of foreign compounds, resulting in facilitating their transport into aqueous compartment of the cell.

17.1.3 Acetaminophen-Induced Hepatocyte Injury

The metabolites of many chemicals are responsible for their toxicities. Hepatic and renal excretions are two major pathways for the elimination of foreign compounds and their metabolites from the body. Many chemicals are activated to toxic metabolites through biotransformation catalyzed by hepatic enzymes, while glutathione S-transferase acts as a vital liver detoxification enzyme.

For examples, the toxicity and carcinogenicity of formaldehyde has been attributed to its ability to form adducts with DNA and proteins. Increased formaldehyde levels are correlated with protein carbonylation in hepatocytes. Drug-induced liver injury can cause various forms of acute and chronic liver disease. Moreover, acetaminophen is a common cause of acute drug-induced liver failure. The role of alcohol consumption as a risk factor in acetaminophen injury to the liver was also investigated.

The metabolite of acetaminophen is a reactive intermediate which is capable of binding to cellular molecules, causing liver cell damage. Ceramide was found to exacerbate cells injury with GSTA1 mRNA level reducing. In contrast, administration of oltipraz alleviated cells damage by acetaminophen-induced hepatocyte injury, and cause a significant decrease in hepatocyte nuclear factor 1 and glutathione S- transferase expressions.

17.1.4 Other Factors

Research aimed at elucidating the molecular mechanism of the pathogenesis of induced liver diseases is essential for preventing liver injuries. Besides genetic factors that determine the susceptibility of specific individuals to chemical-induced liver injuries, environments, and lifestyles also play significant roles. As mentioned above, excessive alcohol consumption is a risk factor for intrinsic hepatotoxicity from acetaminophen, which is the leading cause of drug overdose and acute liver failure.

In addition, the hydrolytic metabolites of flutamide and phenacetin also appear to be associated with hepatotoxicity. Glutathione polymorphisms are also believed to exert a critical role in cellular protection against oxidative stress and toxic foreign chemicals. They involve in detoxification of a variety of electrophilic compounds, including oxidized lipid, DNA, and catechol products generated by reactive oxygen species–induced damage to intracellular molecules.

17.2 Metabolic Intermediate–Associated Kidney Toxicities

The kidney is the most important organ for excreting foreign compounds including foods, chemicals, drugs and their metabolites. This organ is efficient in the elimination of toxicants from the body and is critical in the body's defense against foreign compounds. Besides, renal excretion also plays an important role in eliminating conjugation products resulting from phase II detoxification reactions. Renal excretion involves glomerular filtration, active tubular secretion, and passive tubular readsorption.

Organic cations or anions that are secreted by the kidney may be either hydrophobic or hydrophilic. Structurally diversified organic cations and anions, including positively and negatively charged drugs and their metabolites, are secreted in the proximal tubule of the kidney. Organ-specific toxicity of several halogenated alkenes to the kidney has also been elucidated.

Haloalkene-derived cysteine S-conjugates may be substrates for renal cysteine conjugate beta-lyases. The formation of reactive intermediates by cysteine conjugate beta-lyase may play a role in the target-organ toxicity and the possible renal tumorigenicity of several chlorinated olefins. Conjugation of halogenated alkenes with glutathione, catalyzed by hepatic glutathione S-transferases results in the formation of glutathione S-conjugates. The ability of the kidney to concentrate glutathione and cysteine S-conjugates is responsible for its carcinogenicity.

Several classes of compounds are converted by glutathione conjugate formation to toxic metabolites, such as those derived from halogenated alkanes, alkenes, and quinones. Selective toxicity to the kidney is due to its capability to accumulate intermediates and to bioactivate these intermediates to toxic metabolites. Their covalent binding to cellular macromolecules is responsible for the toxicity of the parent compounds.

Conjugation of halogenated alkenes with glutathione (GSH), catalyzed by hepatic GSH S-transferases, results in the formation of glutathione S-conjugates. The ability of the kidney to concentrate GSH and cysteine S-conjugates is responsible for its carcinogenicity. Moreover, hepatic glutathione S-conjugate forms followed by cleavage to the corresponding cysteine S-conjugates.

Evidence has been accumulating that several classes of compounds are converted by glutathione conjugate formation to toxic metabolites, such as those derived from halogenated alkanes, alkenes and quinones. Selective toxicity to the kidney is due to its capability to accumulate intermediates and to bioactivate these intermediates to toxic metabolites.

17.3 Metabolite–Associated Cancer and Other Toxicities

When production of reactive oxygen species and antioxidative defense is imbalanced, the state of oxidative stress occurs. Reactive oxygen species formed during oxidative stress are not only cytotoxic and can damage protein, nucleic acid, and lipid functions but also can modulate signal transduction in cells and disease conditions. CYP450 enzymes consist of a large group of heme-containing proteins, which are involved in metabolism of foreign compounds and convert them into derivatives that can be eliminated from the body followed by conjugation reactions. During these processes, CYP450 metabolism results in production of activated metabolites that can cause gene mutations. Reactive oxygen stress plays crucial role in biological homeostasis and pathogenesis of human diseases including cancer.

17.3.1 Cancer

Reactive oxygen species (ROS) have a dual role in cell metabolism. At low levels, ROS act as signal transducers to activate cell proliferation and migration. In contrast, at high levels, ROS cause damage to cells and organelles. Cancer cells were found to exhibit higher basal levels of ROS in comparison with normal cells. Many of the phenotypes of cancer cells can be the result of mutations causing DNA damages.

For examples, enzymes glucuronic transferase and glutathione S-transferase metabolize to detoxify foreign compounds. The glutathione S-transferase (GST) supergene family is an important part of cellular enzymic defense against endogenous and exogenous chemicals. Associations between GSTM1 and GSTT1 genotypes and their risk in diseases have been observed in some studies of colon cancers.

17.3.2 Neurodegeneration

Oxidative stress affects biomolecules (proteins, lipids, and DNA), which can also lead to neuronal dysfunctions. Alzheimer and Parkinson diseases are the two most common age-related diseases characterized by neurodegeneration. Oxidative stress is linked to the development of neural dysfunction, which suggests a pathogenic role of oxidative stress in Alzheimer and Parkinson diseases.

17.3.3 Cataract

The pathogenesis of cataract is influenced by a number of factors including oxidative stress. Studies were carried out to determine the risk of genetic polymorphisms of GST isoforms for developing of age-related cataracts. GSTT1 and GSTM1 deletion genotypes are associated with a variety of pathological conditions of ophthalmic diseases. Such genotypes may confer risk for the development of cataracts.

17.4 Drug Efficacy and Adverse Responses

Most drugs are hydrophobic substances that are able to entry through lipid bilayers into cells, where drugs interact with their target receptors or proteins. Xenobiotic-metabolizing enzymes convert drugs and foreign compounds into hydrophilic derivatives that are more easily eliminated through excretion into the aqueous compartments of the tissues.

Many therapeutic agents have been reported to initiate adverse drug response. Xenobiotic-metabolic enzymes play a significant role in adverse drug response, because many of drugs that are associated with adverse response are subjected to metabolism by xenobiotic-metabolic enzymes, particularly CYP450 isoenzymes.

Electrophilic metabolite is involved in the toxication of numerous drugs. Formation of many electrophilic metabolites is catalyzed by CYP450s. To elucidate adverse response of a drug, it is essential to understand drug metabolism and the enzymes involved in order to identify the metabolite generated in the metabolism and to determine the enzymes involved in drug detoxification.

17.4.1 Drug Efficacy

As mentioned above, xenobiotic metabolizing enzymes are critical to the metabolism of drugs and other xenobiotics. The extent of metabolism affects the efficacy and the adverse response of a drug. When the enzymes are excessively active, a drug is metabolized too quickly, resulting in a quick loss of its therapeutic efficacy. In contrast, when the enzymes are overly inactive, a drug is metabolized too slowly, leading to its accumulation in the bloodstream and resulting in a decrease in the plasma clearance of drug.

Electrophilic metabolite is involved in the toxication of numerous chemicals. To elucidate adverse response of a drug, it is essential to understand drug metabolism and the enzymes involved. That is, to identify the metabolite generated in metabolism and to determine the enzymes involved in drug detoxification.

17.4.2 Drug Adverse Responses

Phase I activation reactions carry out oxidation, reduction or hydrolysis reactions, leading to the introduction of functional groups, resulting in a modification of compounds with -OH, -COOH, -SH, or -NH2 group. The addition of such functional groups increases the solubility of drugs and serves as substrates for the phase II conjugating enzymes.

CYP450s are the most actively studied xenobiotic-metabolizing enzymes, since they are responsible for metabolizing the vast majority of therapeutic drugs, like some synthetic drugs undergoing metabolic activation to form reactive metabolites that are often associated with drug toxicity. Some herbal components may also be converted to toxic, or even mutagenetic and carcinogenic, metabolites by CYP450s.

Like some synthetic drugs undergoing metabolic activation to form reactive metabolites which are often associated with drug toxicity, some herbal components may also be converted to toxic, or even mutagenetic and carcinogenic metabolites by CYP450s. Moreover, some naturally occurring flavonoids, such as quercetin, can undergo metabolic activation, resulting in reactive intermediates capable of forming DNA adducts and genotoxicity. Many therapeutic agents have been reported to initiate adverse drug response, in which xenobiotic-metabolic enzymes play a significant role.

Among esterase, carboxylesterases are well known to be involved in the hydrolysis of a variety of drugs, and arylacetamide deacetylase is involved in the hydrolysis of flutamide and phenacetin. While esterase is involved in drug metabolism with different substrate specificity. Their studies promote our understanding of clinical pharmacotherapy and drug development.

17.5 Chemoprevention Against Toxicities

Metabolizing activation enzymes are found in many tissues in the body with the highest levels located in liver and intestines. They can convert certain chemicals to highly reactive toxic and carcinogenic metabolites. Chemicals that can be converted by xenobiotic metabolism to cancer-causing derivatives are called carcinogens. Besides, genetic polymorphisms of metabolic enzymes can also increase the susceptibility of organisms to carcinogenesis and inflammatory diseases.

17.5.1 Toxic Metabolites

Xenobiotic-metabolic enzymes produce electrophilic metabolites that can react with nucleophilic cellular macromolecules such as DNA, RNA, and protein, which can cause cell and organ toxicities. Chemicals or drugs that can be converted by xenobiotic metabolism to cancer-causing derivatives are called carcinogens. Reaction of electrophiles with DNA can sometimes result in cancer through mutation of genes. Reaction of electrophiles with DNA can sometimes result in cancer through mutation of genes.

When protective defenses are overwhelmed by excess toxicant insult, the effects of reactive intermediate species lead to deregulation of cell signaling pathways and dysfunction of biomolecules, leading to failure of target organelles and eventual cell dysfunction.

17.5.2 Chemoprevention

A major mechanism of chemical protection against carcinogenesis and mutagenesis mediated by metabolic electrophiles is the induction of enzymes that involve their metabolism, particularly phase II detoxification enzymes, such as glutathione S-transferases (GSTs), uridine diphosphate-glucuronosyltransferases, and NAD (P)H:quinone reductase. In phase II reactions, conjugation reactions usually require the substance to have oxygen, nitrogen, and sulfur atoms that serve as acceptor sites for a hydrophilic moiety.

Phase II detoxification reactions improve water solubility of foreign compounds, facilitate the elimination of drugs, and inactivation of electrophilic toxic metabolites. A number of metabolic enzyme systems important in the protection of cells from chemical-induced oxidative damages, such as glutathione peroxidase, glutathione reductase, catalase, and superoxide dismutase, have been extensively investigated.

17.5.3 Induction of Detoxification Enzymes

Detoxification enzymes, such as glutathione transferase, quinone reductase, and glucuronosyltransferase, are induced by a low concentration of a variety of antioxidant substances, such as oltipraz, dithiolethiones, and isothiocyanates. Oltipraz, a substituted dithiolethione, is a typical example of chemoprevention agent. Its chemopreventive action was found to associate with increases in the activities of phase II detoxification enzymes, such as glutathione S-transferase (GST), which catalyzes the detoxification of aflatoxin.

Oltipraz protects against aflatoxin-induced hepatocarcinogenesis during carcinogen exposure. Such protection reflects an alteration in the metabolism and disposition of aflatoxin induced by oltipraz. Similarly, sulfur-containing substances derived from garlic and onion have also been shown to prevent carcinogenesis.

The chemoprevention effects of organosulfur compounds, S-methyl cysteine, were also examined on hepatotoxicity. S-methyl cysteine was found to significantly decrease the number and area of glutathione S-transferase placental form foci. The results support S-methyl cysteine as chemoprevention agents.

GST activity in the liver was also increased significantly after administration of S-allyl cysteine. These results support the hypothesis that S-allyl cysteine exhibits chemoprevention activity by exerting specific effects on carcinogen detoxification systems. In addition, isothiocyanates found in cruciferous vegetables were reported to have potent cancer-prevention activities. Among the best characterized isothiocyanates is sulforaphane, which can simultaneously modulate multiple cellular targets involved in carcinogenesis, including modulating carcinogen-metabolizing enzymes.

17.5.4 Nrf2- Keap1 Pathway

Nuclear factor erythroid 2-related factor 2 (Nrf2) is a master transcriptional regulator of genes, which defends organisms against oxidative insults and toxicities. It is critical to preserve cell functions and viabilities during oxidative stress. As a transcription factor, Nrf2 regulates various genes involved in redox elements, protein degradation, DNA repair, and xenobiotic metabolism,

Nrf2-Keap1 (Kelch-like ECH-associated protein 1) pathway is one of major signaling cascades involved in cell defense against endogenous and exogenous stress. The mode of Nrf2 activation plays an important role in preventing disease initiation or progression. Consequently, the Nrf2-Keap1 pathway and its target genes provide protection against various disease conditions including tumorigenesis and cancer. The subject of Nrf2- Keap1 pathway is further discussed in another chapter.

Bibliography

Aliyev AT, Panieri E, Stepanic V et al (2021) Involvement of NRF2 in breast cancer and possible Therapeutical role of polyphenols and melatonin. Molecules 26(7):1853
Anders MW, Dekant W (1998) Glutathione-dependent bioactivation of haloalkenes. Annu Rev Pharmacol Toxicol 38:501–537
Bolton MG, Muñoz A, Jacobson LP et al (1993) Transient intervention with oltipraz protects against aflatoxin-induced hepatic tumorigenesis. Cancer Res 53(15):3499–3504

Chang Y, Wang F, Yang Y et al (2019) Acetaminophen-induced hepatocyte injury: C2-ceramide and oltipraz intervention, hepatocyte nuclear factor 1 and glutathione S-transferase A1 changes. J Appl Toxicol 39(12):1640–1650

Chen C-H (2012) Activation and detoxification enzymes: functions and implications. Springer, New York

Chen C-H (2020) Xenobiotic metabolic enzymes: bioactivation and antioxidant defense. Springer Nature, Cham

Dekant W, Vamvakas S, Koob M et al (1990) A mechanism of haloalkene-induced renal carcinogenesis. Environ Health Perspect 88:107–110

Elkashty OA, Tran SD (2021) Sulforaphane as a promising natural molecule for cancer prevention and treatment. Curr Med Sci 41(2):250–269

Fukami T, Yokoi T (2012) The emerging role of human esterases. Drug Metab Pharmacokinet 27(5):466–477

Fukushima S, Takada N, Wanibuchi H et al (2001) Suppression of chemical carcinogenesis by water-soluble organosulfur compounds. J Nutr 131(3s):1049S–1053S

Gonzalez FJ, Gelboin HV (1993) Role of human cytochrome P-450s in risk assessment and susceptibility to environmentally based disease. J Toxicol Environ Health 40(2–3):289–308

Gu X, Manautou JE (2012) Molecular mechanisms underlying chemical liver injury. Expert Rev Mol Med 14:e4

Harman AW, McKenna M, Adamson GM (1990) Postnatal development of enzyme activities associated with protection against oxidative stress in the mouse. Biol Neonate 57(3–4):187–193

Hatono S, Jimenez A, Wargovich MJ (1996) Chemopreventive effect of S-allylcysteine and its relationship to the detoxification enzyme glutathione S-transferase. Carcinogenesis 17(5): 1041–1044

Hayes JD, Strange RC (2000) Glutathione S-transferase polymorphisms and their biological consequences. Pharmacology 61(3):154–166

Jelic MD, Mandic AD, Maricic SM et al (2021) Oxidative stress and its role in cancer. J Cancer Res Ther 17(1):22–28

Kensler TW (1997) Chemoprevention by inducers of carcinogen detoxication enzymes. Environ Health Perspect 105(Suppl 4):965–970

Koob M, Dekant W (1991) Bioactivation of xenobiotics by formation of toxic glutathione conjugates. Chem Biol Interact 77(2):107–136

Kwon H-J, Won Y-S, Park O et al (2014) Aldehyde dehydrogenase 2 deficiency ameliorates alcoholic fatty liver but worsens liver inflammation and fibrosis in mice. Hepatology 60(1): 146–157

Leinonen HM, Kansanen E, Polonen P et al (2014) Role of the Keap1-Nrf2 pathway in cancer. Adv Cancer Res 122:281–320

Li Y, Hao B, Muhammad I et al (2019) Acetaminophen-induced reduction in glutathione-S-transferase A1 in hepatocytes: a role for hepatic nuclear factor 1α and its response element. Biochem Biophys Res Commun 516(1):251–257

Lieber CS (1994) Susceptibility to alcohol-related liver injury. Alcohol Alcohol Suppl 2:315–326

MacAllister SL, Choi J, Dedina L, O'Brien PJ (2011) Metabolic mechanisms of methanol/formaldehyde in isolated rat hepatocytes: carbonyl-metabolizing enzymes versus oxidative stress. Chem Biol Interact 191(1–3):308–314

Nakamura H, Takada K (2021) Reactive oxygen species in cancer: current findings and future directions. Cancer Sci 112(10):3945–3952

Novak D, Lewis JH (2003) Drug-induced liver disease. Curr Opin Gastroenterol 19(3):203–215

Primiano T, Egner PA, Sutter TR et al (1995) Intermittent dosing with oltipraz: relationship between chemoprevention of aflatoxin-induced tumorigenesis and induction of glutathione S-transferases. Cancer Res 55(19):4319–4324

Schmidlin CJ, Shakya A, Dodson M et al (2021) The intricacies of NRF2 regulation in cancer. Semin Cancer Biol 76:110–119

Sid B, Verrax J, Calderon PB (2013) Role of oxidative stress in the pathogenesis of alcohol-induced liver disease. Free Radic Res 47(11):894–904

Sireesha R, Laxmi SGB, Mamata M et al (2012) Total activity of glutathione-S-transferase (GST) and polymorphisms of GSTM1 and GSTT1 genes conferring risk for the development of age related cataracts. Exp Eye Res 98:67–74

Strange RC, Fryer AA (1999) The glutathione S-transferases: influence of polymorphism on cancer susceptibility. IARC Sci Publ 148:231–249

Teng S, Beard K, Pourahmad J et al (2001) The formaldehyde metabolic detoxification enzyme systems and molecular cytotoxic mechanism in isolated rat hepatocytes. Chem Biol Interact 130–132(1–3):285–296

Thanan R, Oikawa S, Hiraku Y et al (2014) Oxidative stress and its significant roles in neurodegenerative diseases and cancer. Int J Mol Sci 16(1):193–217

Wogan GN, Hecht SS, Felton JS et al (2004) Environmental and chemical carcinogenesis. Semin Cancer Biol 14(6):473–486

Yan MH, Wang X, Zhu X (2013) Mitochondrial defects and oxidative stress in Alzheimer disease and Parkinson disease. Free Radic Biol Med 62:90–101

Zhou S-F, Xue CC, Yu X-Q, Wang G (2007) Metabolic activation of herbal and dietary constituents and its clinical and toxicological implications: an update. Curr Drug Metab 8(6):526–553

Zhou J, Sun C, Yang L et al (2022) Liver regeneration and ethanol detoxification: a new link in YAP regulation of ALDH1A1 during alcohol-related hepatocyte damage. FASEB J 36(4): e22224

Metabolic Enzyme Induction for Health Benefits

18

In enzymatic reactions, xenobiotics are initially activated by activation enzymes. The generated reactive metabolites then undergo detoxification enzyme–catalyzed reactions, before they are ready for excretion from the body. Hence, induction or inhibition of metabolic enzymes has a significant impact on the extent of toxicity that reactive intermediates generate during xenobiotics metabolism.

An important feature of xenobiotic metabolic enzymes is their ability to be activated or inhibited by inducer modulation compounds. Modulators are small molecules that exhibit induction or inhibition effects on the activities of activation enzymes and/or detoxification enzymes. The action of such small modulation molecule can bind to the enzyme, either covalently or noncovalently, thereby affecting enzyme structure or conformation, so as to either increase or decrease enzymatic activity.

There are convincing evidences that the consumption of certain dietary ingredients may favorably modulate biotransformation of xenobiotic reactive intermediate–derived toxicities. Inducers of detoxification enzymes act as indirect antioxidants by boosting the body's antioxidant activities. Such inducibility of metabolic enzymes is an important toxicologically relevant feature associated with xenobiotic metabolism.

Having diets rich in vegetables and fruits are found to have a lower risk of developing disease conditions. Accordingly, diets rich in detoxification enzyme inducers have been investigated to promote the scavenge of reactive oxygen species or free radicals. Such metabolic enzyme inducers that act catalytically to detoxify reactive intermediates are regulated by transcription factor Nrf2. As discussed before, the Keap1/Nrf2/ARE pathway defends against xenobiotic reactive intermediate–mediated oxidative stress and its associated inflammation.

18.1 Metabolic Enzyme Modulation

Modulation of metabolic enzymes affects the generation of xenobiotic reactive intermediate. In general, induction of activation enzymes can lead to a higher expression of enzyme activity and result in enhancements of the activation rate and the production of potentially toxic metabolic intermediates.

While, inducers of detoxification enzymes act as indirect antioxidants by boosting the body's antioxidant systems, thus exerting antioxidant activities. Consequently, the induction of detoxification enzymes can lead to a higher expression of detoxification activity and an increase in the rate of detoxifying reactions.

Enzyme modulators are small molecules that exhibit induction or inhibition effects on the activities of activation enzymes or detoxification enzymes. For a foreign compound which toxicity is caused by metabolic activation, the inhibition of activation enzymes would decrease xenobiotic toxic effects, while the induction of detoxification enzymes would also decrease xenobiotic toxicity.

However, it is important to understand that the consequence of enzyme modulation is dependent on not only the specific effect on activation or detoxification enzymes but also the balance between the activities of activation and detoxification enzymes.

18.1.1 Activation Enzyme Modulation

CYP450 isomers are major activation enzymes for xenobiotic metabolism. They play a major role in the biotransformation and the metabolism of foreign compounds. The activity and the potential toxicity of many drugs are strongly influenced by CYP450-catalyzed biotransformation reactions.

Figure 18.1 illustrates how induction and inhibition of activation enzyme may potentially affect xenobiotic metabolism and its related reactive intermediate formation.

(a) Induction of Activation Enzyme

As shown in Fig. 18.1, a significant induction of activation enzyme to a higher activity can result in a potential enhancement in xenobiotic related toxic effect. For example, phenobarbital is an inducer of activation enzyme during drug metabolism. Its induction mainly involves the activation of CYP450 enzyme.

For example, phenobarbital was found to enhance the activation of cocaine, resulting in the potentiation of cocaine–induced hepatotoxicity. Another example is polycyclic aromatic hydrocarbons (PAHs), which are abundant in the environment. Its metabolic activation is carried out by CYP450. PAH such as benzo[a] pyrene shows significant induction of CYP450 isomers. Examination of human hepatoma cells revealed that PAHs affect metabolism of benzo[a]pyrene catalyzed by CYP450 isozymes.

a. Induction of Activation Enzyme

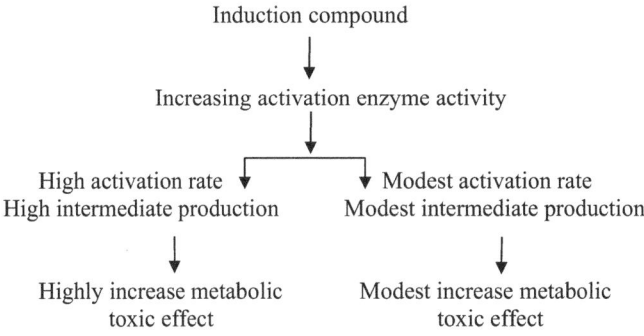

b. Inhibition of Activation Enzyme

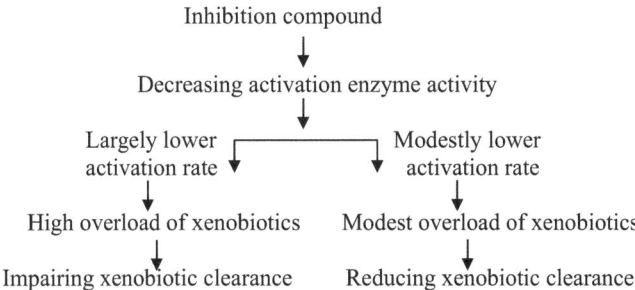

Fig. 18.1 Activation enzyme modulation affecting xenobiotic metabolism

(b) Inhibition of Activation Enzyme

A modest inhibition of CYP450 can lead to a lower activation activity and producing less reactive intermediates. As a result, a reduction in xenobiotic toxic effect may occur. However, an over inhibition of CYP450 activity can lead to an accumulation of foreign compounds, leading to impaired metabolic clearance.

For example, grapefruit juice components (naringin and furanocoumarins) are examples of inhibitors for activation enzyme. The consumption of grape fruit juice with drugs taken orally has been reported to inhibit the activity of intestinal CYP3A4, resulting in a decrease in the metabolism of drugs.

(c) Involvement of Activation Modification

In general, the induction or inhibition of metabolic activation enzyme is an issue of complexity, since it can lead to either increasing or decreasing the toxicity of a foreign compound. It is important to note that the effects of activation enzyme modulation are dependent not only on the resulted activity of activation enzyme, but also on the rate of respective detoxification enzyme.

18.1.2 Detoxification Enzyme Modulation

Induction of detoxification enzymes enhances the rate of detoxification reactions, leading to a faster activity in forming conjugated metabolites, thus facilitating the excretion of xenobiotics. Conversely, inhibition of detoxification enzymes decreases the rate of detoxification reactions, leading to a lower reaction in forming conjugated metabolites, which can lead to an enhancement of xenobiotic toxic effect.

Among detoxification enzymes, UDP-glucuronosyl transferase (UGT), glutathione-S-transferases (GST), and NAD(P)H quinine reductase (NQO) are most studied with respect to enzyme modulation. Figure 18.2 illustrates how induction and inhibition of detoxification enzymes affect xenobiotic toxic effects.

(a) Induction of Detoxification Enzyme

A typical example of detoxification enzyme inducer is oltipraz, a 1,2-dithiole-3-thione derivative. An increase in the expression of detoxification enzyme genes by

a. Induction of Detoxification Enzyme

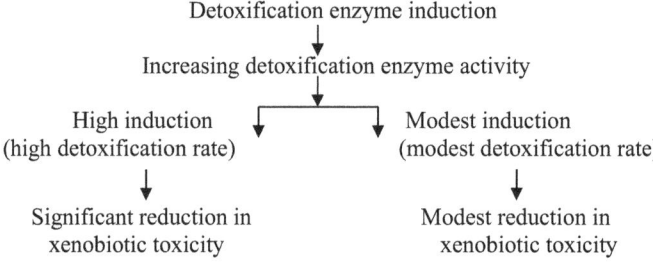

b. Inhibition of Detoxification Enzyme

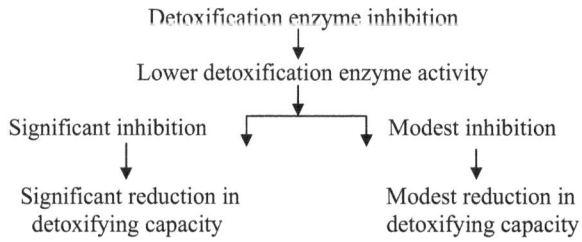

Fig. 18.2 Detoxification enzyme modulation affecting xenobiotic metabolism

oltipraz was found to attribute to a reduction in aflatoxin B1-induced hepatocarcinogenesis. Another example is phenobarbital, which was reported to induce glucuronide conjugation catalyzed by UGT. Glutathione S-transferase (GST) activity detoxifies polycyclic aromatic hydrocarbons. Protein thiol or glutathione (GSH) plays an important role in the induction of GST.

Quinone reductase (QR) is a phase II detoxification enzyme that also plays an important role in detoxifying quinones. Genistein and resveratrol upregulate QR expression in breast cancer cells. Induction of QR may represent an additional mechanism for breast cancer protection.

(b) Inhibition of Detoxification Enzyme

A marked increase in formaldehyde (HCHO) cytotoxicity was observed when either alcohol dehydrogenase or aldehyde dehydrogenase is inhibited. The toxicity and carcinogenicity of HCHO has been attributed to its ability to form adducts with DNA and proteins. Inhibition of GSH-dependent HCHO dehydrogenase activity by prior depletion of GSH was found to markedly increase hepatocyte susceptibility to HCHO.

(c) Less Complications for Induction

Comparison of Fig. 18.2 with Fig. 18.1 reveals that the modulation of detoxification enzyme is less complicated than that of activation enzyme. Induction of detoxification enzymes enhances the detoxification rate, resulting in faster formation of conjugated metabolites and facilitation of xenobiotics excretion. On the other hand, inhibition of detoxification enzymes reduces their detoxifying capacities.

18.1.3 Balance Between Activation and Detoxification Inductions

The efficiency of foreign compound metabolism is dependent on relative activities of activation and detoxification enzymes. In general, a minimal xenobiotic toxicity is found in either a fine balance rate between activation and detoxification expressions or a higher expression of detoxification activity than activation activity.

As the rates of activation and detoxification are balanced, the generated reactive intermediates are present at a minimum, leading to lower xenobiotic toxicity. Meanwhile, when the rate of activation is significantly lower than that of detoxification, reactive intermediates are present at a minimum, which may also result in a lower xenobiotic toxicity.

Consequently, a fine balance between the expressions of activation enzyme and detoxification enzyme is critical to determine whether reactive intermediates are adequately detoxified or may cause toxic effects. In general, the rate of reactive intermediate production comparable with that of detoxification reaction is essential to achieve this fine balance.

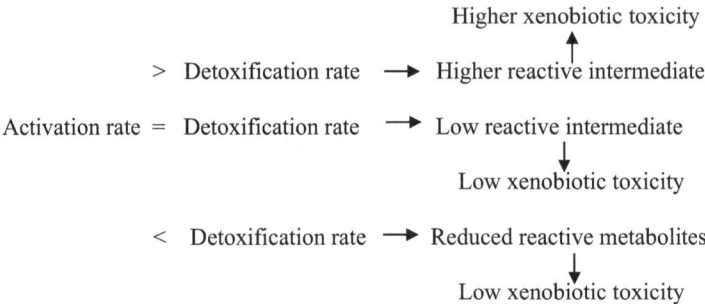

Fig. 18.3 Relative rates of metabolic enzymes affecting xenobiotic toxicity

Accordingly, detoxification metabolism that functions appropriate appears to be an important means of preventing reactive intermediate–mediated toxic effects. For example, induction of CYP450 expression is likely to either increase or decrease the toxic effects of xenobiotics, depending on the relative activity of detoxification enzymes. When the induction of CYP450 results in the rate of activation comparable with that of detoxication, such induction is likely to produce beneficial effects. Nevertheless, if the induction of CYP450 results in the rate of activation being significantly higher than that of detoxification, then the induction may produce harmful effects.

Figure 18.3 summarizes the relative rates of activation and detoxification expressions in relation to xenobiotics-mediated toxicities. The mechanisms underlying the defense against xenobiotics-mediated toxicity include a reduction of metabolic activation by appropriately inhibiting activation enzymes, an enhancement of metabolic detoxification by inducing detoxification enzymes, as well as an increase of antioxidant defense.

18.2 Varieties of Metabolic Enzyme Inducers

Detoxification enzymes, such as glutathione transferase (GTS), NAD(P)H:quinone reductase (QR), and UDP-glucuronosyltransferases (UGS), are induced by a variety of chemical compounds, including phenolic, flavonoids, isothiocyanate, and 1,2-dithiole-3-thiones. For examples, induction of the expressions of QR and GST by phenolic antioxidants has been found to correlate with the chemoprevention properties of phenolic compounds.

Moreover, isoflavone genistein was found to be an inducer of sulfotransferases, since genistein significantly induces mRNA expression of sulfotransferase. On the other hand, ethacrynic acid (a plant phenolic acid) is both an inhibitor and an inducer of GST activity. While comparative studies revealed that a number of coordinated factors are involved in ethacrynic acid–enhanced expression of GST and other detoxification enzymes.

18.2 Varieties of Metabolic Enzyme Inducers

18.2.1 Activation Enzyme Inducers

To exert their toxic effects, most foreign compounds require activation catalyzed by activation enzymes to form electrophilic reactive intermediates or metabolites. The reactive groups are substrates of metabolic enzymes, which are either positively charged or have an atom carrying a partial positive charge.

Electrophilic reactive intermediate or metabolite prefers to interact with a nucleophilic group in the amino acid side chain of metabolic enzymes. Activation enzyme inducers, including polycyclic aromatic hydrocarbons, cannabidiols, phenobarbital, and allyl sulfide, are described below.

(a) Polycyclic Aromatic Hydrocarbons

Polycyclic aromatic hydrocarbons (PAHs) are environmental chemicals. They are activated by metabolic activation enzymes to highly reactive metabolites, which are capable of attacking cellular DNA. CYP450 enzymes are important to the metabolic activation of PAHs to epoxide intermediates, which are then converted by epoxide hydrolase to toxic diol-epoxides and are further metabolized by CYP450s to carcinogenic metabolites such as PAH diol-epoxides or reactive PAH o-quinones.

(b) Cannabidiols

Cannabidiols from cannabis plants were investigated to induce the expression of CYP1A1. Such induction of CYP1A1 expression by cannabidiols was significantly attenuated by reducing the expression of aryl hydrocarbon receptor, a transcription factor that regulates gene expression.

(c) Phenobarbital

In cattle, phenobarbital was found to upregulate drug metabolic enzyme mRNA levels. Phenobarbital increases CYP450 enzyme content and its related enzyme activities. However, contradictory results were obtained for CYP1A, where a decreased catalytic activity was observed.

(d) Allyl Sulfide

Diallyl sulfide and other organosulfur compounds are main constituents of garlic. Diallyl sulfide is an inhibitor of activation enzyme CYP2E1. CYP2E1-mediated alcohol or drug metabolism produces reactive metabolites, which can damage DNA, protein, and lipid membranes, subsequently causing liver damage.

18.2.2 Detoxification Enzyme Inducers

1,2-dithiol-3-thione and several substituted analogs were found to induce quinone reductase activity levels in murine hepatoma cells. Quinone reductase protects cells against the toxicity of quinones by inducing major detoxification enzymes. The induction of glutathione S-transferases (GST) by antioxidants enables to tolerate exposure to carcinogens. GST are also induced by butyrate, a product of plant foods.

Moreover, the induction of detoxication enzymes such as glutathione transferases and quinone reductase by dimethyl fumarate suggests the potential value of this compound as a protective agent against chemical carcinogenesis and electrophile toxicity. Other detoxification enzyme inducers include rosemary extract, sulforamate derivative, fumaric derivative, isothiocyanate, indole-3-carbinol, and 4-bromoflavone. They are briefly discussed below.

(a) Rosemary Extract

The effects of dietary intake of rosemary extract on the liver activities of glutathione-S-transferase and NAD(P)H-quinone reductase were evaluated. Rosemary extract was found to be an effective enhancer of the activities of these enzymes.

(b) Sulforamate Derivative

Several derivatives of sulforamate were synthesized to evaluate their effectiveness as monofunctional inducer of the NAD(P)H quinone oxidoreductase (QR). Their induction potential was found to be comparable to sulforaphane.

(c) Fumaric Derivative

Dimethyl fumarate is a potent inducer of cytosolic NAD(P)H:quinone oxidoreductase activity in hepatoma cells. It also induces quinone reductase. These findings suggest the potential value of fumaric derivative as a protective agent against chemical carcinogenesis and other forms of electrophile toxicity.

(d) Isothiocyanate

Isothiocyanate sulforaphane isolated from cruciferous vegetables is known to induce the activities of detoxification enzymes, including glutathione reductase, glutathione-S-transferase, thioredoxin reductase, and NAD(P)H:quinone oxidoreductase-1, resulting in a cytoprotective action against oxidative damage.

(e) Indole-3-Carbinol

Indole-3-carbinol, a compound present in cruciferous vegetables, increases hepatic and intestinal glutathione S-transferase and epoxide hydrolase activities.

The results indicate that Brassica oleracea also contains other compounds that are responsible for inducing the activities of these enzymes.

(f) 4-Bromoflavone

4'-bromoflavone was found to significantly induce quinone reductase activity and also effectively induce glutathione S-transferase in hepatoma cells. Induction mediated by 4'-bromoflavone is bifunctional (both phase I and phase II enzymes). 4'-bromoflavone is considered as an effective chemoprevention agent.

18.3 Monofunctional and Bifunctional Inducers

Two classes of enzyme inducers in xenobiotic metabolism have been proposed: monofunctional and bifunctional inducers. Monofunctional inducers elevate either the activation or the detoxification enzyme. While bifunctional inducers induce both activation and detoxification enzymes. A direct assay of the activity of quinone reductase in murine hepatoma cells has been utilized to characterize bifunctional and monofunctional inducers. Bifunctional induction was found to be dependent on aryl hydrocarbon receptor function.

In spite of many studies supporting the classification of monofunctional and bifunctional inducers, further investigations are needed to ensure that monofunctional inducers modulate exclusively on either detoxification or activation enzymes, but not both enzymes. Neither monofunctional nor bifunctional inducer seems to display common characteristics in terms of chemical structures or functional groups.

18.3.1 Monofunctional Inducers

The induction of detoxication enzymes by sulforaphane may be a significant component of the anticarcinogenic action of broccoli. Its induction of phase II detoxication enzymes, such as quinone reductase and glutathione S-transferases, affords protection against metabolic-induced toxicity. Like other isothiocyanates, sulforaphane is a monofunctional inducer.

Several derivatives of sulforamate were also evaluated for their effectiveness as monofunctional inducer of NAD(P)H quinone oxidoreductase, a phase II detoxification enzyme. The induction potential of these compounds was found to be comparable to sulforaphane. Moreover, sulforamate with structural similarities to sulforaphane was shown to be a monofunctional inducer of quinone reductase.

Reduction of electrophilic quinones by quinone reductase is an important detoxification pathway. Isoliquiritigenin isolated from tonka bean was found to be a monofunctional inducer of quinone reductase. Dehydroglyasperin isolated from licorice is a potent phase II enzyme inducer. It induces NAD(P)H:oxidoquinone reductase as a monofunctional inducer, but also induces some other phase II detoxification enzymes, such as glutathione S-transferase and glutathione reductase.

18.3.2 Bifunctional Inducers

4'-bromoflavone (4'BF) was found to significantly induce quinone reductase and glutathione S-transferase activities. 4'BF was also found to be a potent inhibitor of CYP450-mediated activity. Thus, induction mediated by 4'BF is bifunctional, involving both phase I and phase II enzymes.

Moreover, oltipraz induces phase II detoxifying enzymes via the antioxidant responsive element, including GST and NAD(P)H:quinone reductase. It also activates phase I metabolism enzyme CYP1A1. Thus, oltipraz is a bifunctional inducer. Similarly, indole-3-carbinol in cruciferous vegetables was also found to be a bifunctional inducer.

Due to significant increase the activities of NAD(P)H-quinone reductase and UDP-glucuronosyl-transferase, juices from cruciferous vegetables were found to protect against the genotoxicity caused by dietary carcinogens. In addition, juice from water cress also acts as a bifunctional inducer by enhancing both phase I and phase II enzymes. Metanil yellow, a food color, is also a bifunctional inducer. It exhibits induction potentials on CYP450-dependent monooxygenases and detoxification enzymes glutathione S-transferase and quinone reductase.

18.4 Inducer–Drug Interactions

Bioavailability is referred to as the degree and rate at which an administered drug is absorbed by the body. Enzyme induction or inhibition that increases or decreases the metabolism of drugs may affect the rate and extent of medications. Knowledge of potent inhibition and induction of drugs can also help minimize the potential adverse drug interactions.

CYP450 enzymes are the major activation enzymes essential for the metabolism of many drugs. CYP450s can be inhibited or induced by drugs, resulting in clinically significant interactions that can cause unanticipated adverse reactions. For instance, in a study of interaction between grapefruit juice and felodipine (a dihydropyridine calcium channel antagonist), grapefruit juice was found to reduce the metabolism of felodipine through selective post-translational down regulation of the expression of CYP3A4 in the intestinal wall.

There are also concerns that antioxidative properties of sulforaphane may interfere with cytotoxic drugs such as doxorubicin. Doxorubicin is a cardiotoxic anticancerous drug that is widely used for a broad range of cancers. Sulforaphane was found to protect the heart from doxorubicin-induced toxicity and damage and enhance the efficacy of doxorubicin in regression of tumor growth. The dosage of doxorubicin required to treat the tumor could be reduced when sulforaphane is administered simultaneously.

Moreover, curcumin is another compound that has been investigated for inducer–drug interaction. Curcumin has potential to be utilized for chemoprevention, due to its antioxidant, anti-inflammatory, and anticarcinogenic properties. Curcumin has been studied for its synergistic effect on cardiomyocytes and cancer cells. Dose-dependent cardiotoxicity of doxorubicin can be significantly minimized by supplementing it with curcumin.

In addition, the regulation of polyamine metabolism in human gastric and colon carcinoma cell lines in response to curcumin was also investigated to provide insight into the polyamine-related mechanisms involved in the cancer cell in response to curcumin and its potential as a chemoprevention agent in the gastrointestinal tract.

Understanding inducer–drug interaction requires further research to explore the potential of utilizing metabolic enzyme inducers for chemoprevention.

Bibliography

Angeloni C, Leoncini E, Malaguti M et al (2009) Modulation of phase II enzymes by sulforaphane: implications for its cardioprotective potential. J Agric Food Chem 57(12):5615–5622

Barry M, Feely J (1990) Enzyme induction and inhibition. Pharmacol Ther 48:71–94

Bianco NR, Chaplin LJ, Monica M, Montano MM (2005) Differential induction of quinone reductase by phytoestrogens and protection against oestrogen-induce DNA damage. Biochem J 385(Pt 1):279–287

Bradfield CA, Bjeldanes LF (1984) Effect of dietary indole-3-carbinol on intestinal and hepatic monooxygenase, glutathione S-transferase and epoxide hydrolase Activities in the rat. Food Chem Toxicol 22(12):977–982

Buetler TM, Gallagher EP, Wang C (1995) Induction of phase I and phase II drugmetabolizing enzyme mRNA, protein, and activity by BHA, ethoxyquin, and oltipraz. Toxicol Appl Pharmacol 135:45–57

Cantiello M, Carletti M, Giantin M et al (2022) Induction by phenobarbital of phase I and II xenobiotic-metabolizing enzymes in bovine liver: an overall catalytic and immunochemical characterization. Int J Mol Sci 23(7):3564

Chen C-H (2012) Activation and detoxification enzymes: functions and implications. Springer Sciences, New York

Chen C-H (2020) Xenobiotic metabolic enzymes: bioactivation and antioxidant defense. Springer Nature, Cham

Conney AH (2003) Enzyme induction and dietary chemicals as approaches to cancer chemoprevention: the Seventh DeWitt S. Goodman Lecture. Cancer Res 63:7005–7031

Das M, Ramchandani S, Upreti RK et al (1997) Metanil yellow: a bifunctional inducer o heptatic phase I and phase II xenobiotic-metabolizing enzymes. Food Chem Toxicol 35(8):835–838

Debersac P, Heydel JM, Amiot MJ et al (2001) Induction of cytochrome P450 and/or detoxication enzymes by various extracts of rosemary: description of specific patterns. Food Chem Toxicol 39(9):907–918

Dinkova-Kostova AT, Talalay P (2008) Direct and indirect antioxidant properties of inducers of cytoprotective proteins. Mol Nutr Food Res 52(Suppl l):S128–S138

Finley JW (2003) The antioxidant responsive element (ARE) may explain the protective effects of cruciferous vegetables on cancer. Nutr Rev 61:250–254

Gerhäuser C, You M, Liu J et al (1997) Cancer chemopreventive potential of sulforamate, a novel analogue of sulforaphane that induces phase 2 drug-metabolizing enzymes. Cancer Res 57(2):272–278

Hodges RE, Minich DM (2015) Modulation of metabolic detoxification pathways using foods and food-derived components: a scientific review with clinical application. J Nutr Metab 2015: Article ID 760689

Kensler TW, Curphey TJ, Maxiutenko Y et al (2000) Chemoprotection by organosulfur inducer of phase 2 enzymes: dithiolethiones and dithiins. Drug Metabol Drug Interact 17(1–4):3–22

Keum YS, Jeong WS, Kong AN (2005) Chemopreventive functions of isothiocyanates. Drug News Perspect 18(7):445–451

Kwak MK, Itoh K, Yamamoto M et al (2002) Enhanced expression of the transcription factor Nrf2 by cancer chemopreventive agents: role of antioxidant response element-like sequences in the nrf2 promoter. Mol Cell Biol 22(9):2883–2892

Lampe JW (2007) Diet, genetic polymorphisms, detoxification, and health risks. Altern Ther Health Med 13(2):S108–S111

Leoncini E, Malaguti M, Angeloni C et al (2011) Cruciferous vegetable phytochemical sulforaphane affects phase II enzyme expression and activity in rat cardiomyocytes through modulation of Akt signaling pathway. J Food Sci 76(7):H175–H181

Lhoste EF, Gloux K, De Waziers I, Garrido S et al (2004) The activities of several detoxication enzymes are differentially induced by juices of garden cress, water cress and mustard in human HepG2 cells. Chem Biol Interact 150(3):211–219

Manson MM, Ball HW, Barrett MC et al (1997) Mechanism of action of dietary chemoprotective agents in rat liver: induction of phase I and II drug metabolizing enzymes and aflatoxin B I metabolism. Carcinogenesis 18:1729–1738

Metere A, Giacomelli L (2017) Absorption, metabolism and protective role of fruits and vegetables polyphenols against gastric cancer. Eur Rev Med Pharmacol Sci 21(24):5850–5858

Miao W, Hu L, Kandouz M et al (2003) Oltipraz is a bifunctional inducer activating both phase I and phase II drug-metabolizing enzymes via the xenobiotic responsive element. Mol Pharmacol 64(2):346–354

Misaki K, Matsui S, Matsuda T (2007) Metabolic enzyme induction by HepG2 cells exposed to oxygenated and nonoxygenated polycyclic aromatic hydrocarbons. Chem Res Toxicol 20(2): 277–283

Moriarty RM, Naithani R, Kosmeder J (2006) Cancer chemopreventive activity of sulforamate derivatives. Eur J Med Chem 41(1):121–124

Nakamura Y, Miyamoto M, Murakami A et al (2003) A phase II detoxification enzyme inducer from lemongrass: identification of citral and involvement of electrophilic reaction in the enzyme induction. Biochem Biophys Res Commun 302(3):593–600

Nasrin S, Watson CJW, Bardhi K et al (2021) Inhibition of UDP-glucuronosyltransferase enzymes by major cannabinoids and their metabolites. Drug Metab Dispos 49(12):1081–1089

Noratto GD, Chew BP, Atienza LM (2017) Red raspberry (Rubus idaeus L.) intake decreases oxidative stress in obese diabetic (db/db) mice. Food Chem 227:305–314

Okey AB, Roberts EA, Harper PA et al (1986) Induction of drug-metabolizing enzymes: mechanisms and consequences. Clin Biochem 19:132–141

Osawa T, Kato Y (2005) Protective role of antioxidative food factors in oxidative stress caused by hyperglycemia. Ann N Y Acad Sci 1043:440–451

Pool-Zobel B, Veeriah S, Böhmer F-D (2005) Modulation of xenobiotic metabolizing enzymes by anticarcinogens – focus on glutathione S-transferases and their role as targets of dietary chemoprevention in colorectal carcinogenesis. Mutat Res 591(1–2):74–92

Prochaska HJ, Talalay P (1998) Regulatory mechanisms of monofunctional and bifunctional anticarcinogenic enzyme inducers in murine liver. Cancer Res 48:4776–4782

Prochaska HJ, Santamaria AB, Talalay P (1992) Rapid detection of inducers of enzymes that protect against carcinogens. Proc Natl Acad Sci 89:2394–2398

Rao PSS, Midde NM, Miller DD et al (2015) Diallyl sulfide: potential use in novel therapeutic interventions in alcohol, drugs, and disease mediated cellular toxicity by targeting cytochrome P450 2E1. Curr Drug Metab 16(6):486–503

Razis AFA, Konsue N, Ioannides C (2018) Isothiocyanates and xenobiotic detoxification. Mol Nutr Food Res 62(18):e1700916

Rekka EA, Kourounakis PN, Pantelidou M (2019) Xenobiotic metabolizing enzymes: impact on pathologic conditions, drug interactions and drug design. Curr Top Med Chem 19(4):276–291

Scapagnini G, Caruso C, Calabrese V (2010) Therapeutic potential of dietary polyphenols against brain ageing and neurodegenerative disorders. Adv Exp Med Biol 698:27–35

Shih PH, Yeh CT, Yen GC (2007) Anthocyanins induce the activation of phase II enzymes through the antioxidant response element pathway against oxidative stress-induced apoptosis. J Agric Food Chem 55:9427–9435

Shimada T (2006) Xenobiotic-metabolizing enzymes involved in activation and detoxification of carcinogenic polycyclic aromatic hydrocarbons. Drug Metab Pharmacokinet 21(4):257–276

Singletary KW (1996) Rosemary extract and carnosol stimulate rat liver glutathione-S-transferase and quinone reductase activities. Cancer Lett 100(1–2):139–144

Song LL, Kosmeder JW 2nd, Lee SK et al (1999) Cancer chemopreventive activity mediated by 4′-bromoflavone, a potent inducer of phase II detoxification enzymes. Cancer Res 59(3): 578–585

Spencer SR, Wilczak CA, Talalay P (1990) Induction of glutathione transferases and NAD(P)H: quinone reductase by fumaric acid derivatives in rodent cells and tissues. Cancer Res 50(24): 7871–7875

Talalay P, Dinkova-Kostova AT, Holtzclaw WD (2003) Importance of phase 2 gene regulation in protection against electrophile and reactive oxygen toxicity and carcinogenesis. Adv Enzym Regul 43:121–134

Teng S, Beard K, Pourahmad J et al (2001) The formaldehyde metabolic detoxification enzyme systems and molecular cytotoxic mechanism in isolated rat hepatocytes. Chem Biol Interact 130–132(1–3):285–296

Tirmenstein MA, Nicholls-Grzemski FA, Zhang JG, Fariss MW (2000) Glutathione depletion and the production of reactive oxygen species in isolated hepatocyte suspensions. Chem Biol Interact 127(3):201–217

Zevin S, Benowitz NL (1999) Drug interactions with tobacco smoking. An update. Clin Pharmacokinet 36:425–438

Zhang Y, Talalay P, Cho CG et al (1992) A major inducer of anticarcinogenic protective enzymes from broccoli: isolation and elucidation of structure. Proc Natl Acad Sci 89(6):2399–2403

Diet Effects on Metabolic Enzymes   19

In the metabolism of foreign compounds, activation enzymes catalyze the functionalization of chemicals to generate reactive intermediates or metabolites, such as heterocyclic aromatic amines and polycyclic aromatic hydrocarbons, which are formed in meat grilled at high temperatures. Such metabolic activation reaction is followed by detoxification enzyme-catalyzed conjugation or non-conjugation reactions to detoxify reactive intermediates or metabolites so as to facilitate their excretion from the body.

Oxidative stress and inflammation mediated by reactive chemical intermediates or metabolites, including reactive oxygen species and free radicals, are involved in the development of disease conditions. To combat such health effects, one main strategy is focused on exogenous antioxidants. Another promising approach is to identify phytochemicals derived from vegetables, fruits, herbs, beverages, and algae to target body's cytoprotective mechanisms. In the case of the latter approach, it is primarily concentrated on the induction of detoxification enzyme expressions involved in metabolic detoxification mechanisms.

With respect to the issue of metabolic detoxification enzyme induction, it is also important to consider genetic polymorphisms of these enzymes, since they account in part for individual variations in health benefits. Another important factor that requires attention is the inducer–drug interactions. Phytochemicals present in vegetables and fruits may influence the pharmacological activities of drugs by modifying their absorption and metabolic characteristics.

Phytochemicals are thought to induce metabolic detoxification enzymes through nuclear factor E2-related factor 2 (Nrf2)-antioxidant response element (ARE) pathway. Nrf2 binding to ARE in the regulatory region of detoxification genes is essential for the induction of detoxification enzymes. Accordingly, activation of the transcription factor Nrf2 has been proposed as a promising strategy in attenuating oxidative stress and its related inflammation.

19.1 Dietary Modulation of Metabolic Enzymes

Extensive animal, cell culture, and epidemiological studies have demonstrated that consumption of vegetables, fruits, and other dietaries is inversely correlated with chronic disease incidences. Besides being the sources of fiber, vitamins, and minerals, vegetables and fruits also contain non-nutritive components, such as phytochemicals, that can provide substantial health benefits beyond basic nutrition.

Human clinical trials have been carried out to elucidate phytochemicals present in vegetables and fruits that demonstrate health benefits. Their non-nutritive components that protect humans against the risk of degenerative and chronic diseases are believed to be linked to the induction of metabolic detoxification enzyme expressions. Accordingly, dietary inducers of metabolic enzymes that are present in vegetables, fruits, herbs, beverages, and algae are discussed below:.

19.1.1 Vegetables

The consumption of vegetables has long been reported to be associated with a reduced risk in the occurrence of degenerative and chronic diseases, such as cardiovascular disease and certain cancers. Among the phytochemicals present in cruciferous vegetables, allium and root vegetables are most investigated for their cytoprotective effects against cytotoxicity and carcinogenesis.

Cruciferous vegetables are vegetables of the family Brassicaceae, including broccoli, cauliflower, Brussels sprouts, cabbage, radish, and kate, while allium vegetables are high in organosulfur compounds with distinctive flavor and smell, such as onions and garlic.

The effectiveness of cruciferous, allium, and root vegetables impact on human health and cell protection against cytotoxicity and carcinogenesis is dependent on the activities of detoxification enzymes encoded by polymorphic genes. Hence, researchers also take into consideration the impact of metabolic enzyme polymorphisms in exploring the health effects of these vegetables.

19.1.2 Cruciferous Vegetables

Cruciferous vegetables that provide protection to living cells against xenobiotic reactive intermediate toxicants include sulfur-containing compounds such as isothiocyanates, sulforaphane, phenethyl isothiocyanate, and 1,2-dithiole-3-thiones, as well as non-sulfur-containing compounds such as indole-3-carbinol. Modulation of the metabolism of tobacco-specific carcinogenic nitrosamines by inhibiting its CYP450-mediated bioactivation was reported. Moreover, potent inducers of detoxification enzymes, such as glutathione S-transferase, UDP-glucuronosyltransferase, and quinone reductase, were also studied. Accordingly, these metabolic enzyme modulators are described below:

19.1 Dietary Modulation of Metabolic Enzymes

(a) Isothiocyanates

Isothiocyanates including sulforaphane and phenethyl isothiocyanate are broadly distributed among cruciferous vegetables, such as cabbages, broccoli, and Brussels sprouts. Isothiocyanates have a chemical structure of R-N=C=S (where R is aliphatic or aromatic group). Natural isothiocyanates are initially present in cruciferous vegetables as glucosinolates, which are converted to isothiocyanates by the enzyme myrosinase.

Cruciferous vegetables are commonly processed before consumption, where significant alteration in their phytochemical compositions may occur. For example, heat processing of Brussels sprouts largely reduces the concentrations of isothiocyanates, so as to affect their ability to induce detoxification enzymes.

Moreover, genetic variations and polymorphisms of detoxification enzymes can also affect the effectiveness of cruciferous vegetables on detoxifying reactive chemical intermediates or metabolites generated by foreign compound metabolism.

Extensive research supports the evidence for the modulation of the mutagenicity of aromatic amines by isothiocyanates. Moreover, epidemiologic data also support the association of high intake of vegetables with low risk of degeneration and chronic disease conditions. Experimental dietary studies in humans have further shown the capacity of cruciferous vegetable constituents to modulate potential disease-preventive mechanisms. Such investigations were primarily carried out on animals and cell or tissue culture studies.

(b) Sulforaphane

Sulforaphane, the most extensively studied isothiocyanate, is an aliphatic isothiocyanate with a chemical formula of -SO-(CH2)4-N=C=S. Sulforaphane is commonly found in cruciferous vegetables, especially broccoli. Broccoli accumulates a significant amount of glucoraphanin, the predominant aliphatic glucosinolate, which is then metabolized to biologically active sulforaphane.

Exploration into broccoli's impact on human health and cell protection against cytotoxicity and carcinogenesis has been actively pursued. Such studies support the association of sulforaphane with the induction of metabolic detoxification enzymes, such as glutathione S-transferases and quinone reductase.

(c) Phenethyl Isothiocyanate

Another studied isothiocyanate is phenethyl isothiocyanate (PEITC). PEITC is an aromatic isothiocyanate, which is primarily metabolized by glutathione conjugation and is excreted in the urine and bile as mercapturate. PEITC affects the metabolism of foreign compounds through modulation of both activation and detoxification enzymes.

Table 19.1 Dietary inducers of detoxification enzymes

Dietary sources vegetables	Induction compounds
Cruciferous vegetable	Sulfur–containing compounds (isothiocyanate, phenethyl–isothiocyanate, sulforaphane, and 1,2-dithiole-3-thione)
	Non-sulfur-containing compound (indole-3-carbinol)
Allium vegetables	Organosulfur compounds:
	Thiols and diallyl sulfides
Root vegetables	Flavonoids, carotenoids, and curcumin
Fruits	Polyphenols (quercetin, kaempferol, pomegranate, anthocyanin, and procyanidin)
	Triterpenes (limonoids and olive oil)
Herb	Ginseng
	Herb–drug interaction
Beverage	Polyphenols: epigallocatechin-3-gallate
Alcohol	Acetaldehyde, curcumin
Algae	Chlorophyll, unsaturated fatty acids, Polysaccharides

(d) 1,2-Dithiole-3-Thione

1,2-dithiole-3-thione is a cyclic sulfur-containing compound present in cruciferous vegetables. Existing evidence has demonstrated that 1,2-dithiole-3-thione is able to efficiently induce the expression of metabolic detoxification enzymes and antioxidants. The derivatives of 1,2-dithiole-3-thione were also reported to decrease the incidence and multiplicity of tumors in animals exposed to chemical carcinogens.

These findings support the ability of 1,2-dithiole-3-thione and its derivatives to enhance the expression of detoxification enzymes. As a potent enzyme inducer and an anticarcinogen agent, 1,2-dithiole-3-thione was also reported to potently upregulate antioxidant genes in cells by stimulating Nrf2 activity.

(e) Indole-3-Carbinol

Unlike isothiocyanates and 1,2-dithiole-3-thione, indole-3-carbinol is a non-sulfur-containing compound and is a hydrolysis product of glucosinolate. Indole-3-carbinol and its dimeric product, 3,3'-diindolylmethane, were found to upregulate the expression of detoxification enzymes, such as glutathione S-transferase.

Dietary inducers of detoxification enzymes, including vegetables, fruits, herbs, beverages, alcohol, and algae, are included in Table 19.1. The table also contains cruciferous vegetables with both sulfur-containing and non-sulfur-containing inducers of detoxification enzymes.

19.1.3 Allium Vegetables

Vegetables, such as shallots, leeks, and chives, are members of the allium family. Organosulfur-containing compounds present in allium vegetables include thiols, disulfides, and trisulfides. A thiol contains a -SH functional group, while a disulfide contains a -S-S- linkage. They are present particularly in garlic, onion, leek, and scallion. Diallyl disulfide is an inhibitor of CYP2E1 enzyme that catalyzes the metabolism of alcohol and drugs.

(a) Thiols

S-alkyl cysteine sulfoxides are typically found in the Alliaceae family. Onion and garlic extracts were shown to induce metabolic detoxification enzymes, such as glutathione-S-transferases and NAD(P)H:oxidoreductase. They have been traditionally used for treating various pathologic conditions. Allium-derived sulfur compounds were also found to play a role in cardiovascular protection.

(b) Diallyl Sulfides

Diallyl sulfide, diallyl disulfide, and diallyl trisulfide derived from garlic and onion were found to increase the activities of detoxification enzymes, such as quinone reductase and glutathione transferase. Epidemiological studies indicate that dietary intake of Allium vegetables decreases the risk of cancer in humans, owing to their ability to induce the activities of metabolic detoxification enzymes. Inducers derived from Allium vegetable, including thiols and diallyl sulfides, are also shown in Table 19.1.

19.1.4 Root Vegetables

Root vegetables, including radish, carrot, and ginger, provide most of the flavonoids, carotenoids, and curcumin present in the human diet. There has been a growing interest in these chemical compounds due to their potential health benefits. Inducers derived from root vegetables, including flavonoids, carotenoids, and curcumin are also presented in Table 19.1.

Flavonoids are the most common group of phenolic compounds in these vegetables. Phenolics are compounds containing at least one aromatic ring and one OH group, and while carotenoids are colored compounds (yellow, orange, or red). Carotenoids can be divided into xanthophylls and carotenes, depending on the presence or absence of oxygen.

Xanthophylls include lutein and zeaxanthin. Carotenes are such as lycopene. Curcumin is the primary curcuminoid in turmeric. Turmeric curcumin is believed to exhibit anti-inflammatory and antioxidant properties.

(a) Flavonoids

Flavonoids are the most important group of phenolics in diets and consist of catechin, anthocyanidin, flavone, and flavanol. Their modulation of metabolic detoxification enzymes, such as UDP-glucuronosyltransferase, glutathione S-transferase, and quinone reductase, is considered a major mechanism of anticarcinogenic effects of flavonoids.

Soybeans are abundant in isoflavones, which are known inducers of metabolic detoxification enzymes. Flavonoid-containing soy was found to enhance the activities of UDP-glucuronosyltransferase and quinone reductase, thus offering protective health effects from potentially toxic reactive intermediate species derived from foreign compound metabolism.

Quercetin was shown to increase the transcription of NADH:quinone oxidoreductase. Research findings also suggest that flavonoids stimulate the transcription of detoxification enzymes, potentially through Nrf2-ARE dependent pathway. Studies have also revealed that dietary flavonoid-mediated induction of intestinal UDP-glucuronosyltransferase may be important for the detoxification of colon carcinogens.

Dietary phenolic derivatives of flavones and isoflavones were also found to be associated with reduced cancer rates. Research evidence also revealed that flavonoids, such as fisetin, galangin, quercetin, kaempferol, and genistein, might offer important chemoprevention from sulfation-induced carcinogenesis.

(b) Carotenoids

Carotenoids are pigments responsible for the red, yellow, or orange color in many vegetables including carrots, tomatoes, and berries. Through acting as antioxidants, carotenoid compounds, such as lycopene, lutein, and zeaxanthin, are known to offer cytoprotective health benefits for the body against cellular damages.

Among carotenoids, lycopene is known as one of the most effective quenchers of reactive oxygen species. Lutein and zeaxanthin have been shown to improve visual acuity and to slow the progression of age-related macular degeneration. Research evidence suggest that carotenoids may upregulate the expressions of metabolic detoxification enzymes and antioxidants through the activation of Nrf2-dependent pathway.

(c) Curcumin

Turmeric is part of the ginger family, and curcumin is the primary curcuminoid in turmeric. Curcumin was reported to induce the expression of glutathione S-transferase. Such induction of detoxification enzymes was considered the potential mechanism underlying the anti-inflammatory and anti-cancer actions of curcumin.

By inducing heme oxygenase 1 and other detoxifying enzymes in neurons, curcumin was also found to protect neurons against various modes of oxidative stress. Epidemiological studies suggested to take curcumin as a preventive agent against aging and neurodegenerative disorders such as Alzheimer's. However, overdoses of curcumin were found to cause toxic effects.

Table 19.1 also includes thiols and diallyl sulfides in allium vegetables and flavonoids, carotenoids, and curcumin in root vegetables.

19.2 Fruits

Cytoprotective properties of fruits, such as citrus fruits, grapes, berries, and pomegranate, have been actively investigated. Polyphenols and triterpenes are the most investigated phytochemicals in fruits for their protective effects against cytotoxicity and carcinogenesis. Polyphenols include pomegranate, anthocyanins, and procyanidins, while triterpenes are a group of chemical compounds consisting of three terpenes. They are a large group of unsaturated hydrocarbons found in the oil of citrus trees. The main triterpenes present in olive oil are oleanolic acid, maslinic acid, and uvaol.

The inducers of detoxification enzymes in fruits, such as polyphenols, eriodyctiol, and quercetin. Kaempferol, pomegranate, anthocyanin, procyanidin, and triterpenes are also listed in Table 19.1. They are described below:

19.2.1 Polyphenols

The most common group of fruit polyphenolic compounds is flavonoids. Flavonoids in citrus fruits, berries, and grapes are such as eriodyctiol, quercetin, kaempferol, pomegranate, anthocyanin, and procyanidin. Phenolic compounds in fruits were reported to exhibit protective effects on the gastrointestinal tract by enhancing the detoxification of reactive intermediates and metabolites, thus preventing foreign compound metabolism-mediated damages to cellular components including proteins, lipids, and nucleic acids.

19.2.2 Eriodyctiol and Quercetin

Among citrus fruits, eriodyctiol is most investigated for its cytoprotective effects on anti-inflammatory and antioxidation. Eriodyctiol's upregulation of the detoxification enzyme heme oxygenase-1 was reported to be beneficial for preventing cardiovascular diseases.

While quercetin has diverse biological properties, including cytoprotective and antitumorigenic effects, an increase in NAD(P)H:quinone oxidoreductase activity in response to quercetin suggests that polyphenols can stimulate transcription of metabolic detoxification systems potentially through ARE-dependent mechanism.

19.2.3 Kaempferol and Pomegranate

Kaempferol, another polyphenol found in fruits, was reported to exhibit pharmacological activities, including antioxidant and anti-inflammatory. Epidemiological studies have found a positive correlation between the dietary consumption containing kaempferol and a reduced risk of developing several disorders, such as cardiovascular diseases and cancers.

Pomegranate juice and peel are also rich in polyphenols. Pomegranate is gaining attention due to its powerful antioxidant properties. Mechanistic studies revealed that pomegranate elevates gene expression of hepatic detoxification enzymes and antioxidants.

Research has also provided substantial evidence to propose that pomegranate-mediated chemoprevention of hepatocarcinogenesis is associated with Nrf2-regulated antioxidant mechanisms.

19.2.4 Anthocyanin and Procyanidin

Anthocyanins are also a type of flavonoids. Blueberry anthocyanins offer cytoprotective effects due to their abilities to inhibit oxidative stress and cell proliferation. Effects of anthocyanin fractions from cultivars of blueberries on detoxification enzymes, such as glutathione-S-transferase and quinone reductase, are the potential mechanism through which anthocyanins exhibit cytoprotective effects, while anthocyanins in grapes were also found to have chemoprevention potential, in part due to their capacity to block carcinogen-DNA adduct formation and to modulate the activities of metabolic enzymes.

Procyanidins are members of the tannin class of flavonoids. They were found to significantly induce the expression of antioxidants such as catalase and superoxide dismutase, as well as the activities of metabolic detoxification enzymes such as quinone oxidoreductase and glutathione peroxidase. Procyanidins from wild grape seeds were used as a potential chemoprevention agent that induces metabolic detoxification enzymes through the Nrf2-ARE pathway.

19.2.5 Triterpenes

Research on the bioactive properties of phytochemicals revealed that the triterpenes present in olive oil are a natural source of antioxidants that could be beneficial for preventing diseases related to oxidative stress damages.

Citrus limonoids are highly oxidized triterpenes, which may provide substantial anticancer activity. It has been reported that consumption of citrus fruits may gain health benefits due to the induction of metabolic detoxification enzymes, such as glutathione S-transferase and NAD(P)H: quinone reductase.

19.3 Herbs

Wild ginseng is a well-known medicinal herb. The protective effects of the root of wild ginseng on benzo[a]pyrene-induced hepatotoxicity were reported. While, co-administration of herbal medicines with therapeutic drugs may cause herb–drug interactions, resulting in low drug efficacy or toxic reactions, such a concern requires attention when taking ginseng.

19.3.1 Ginseng

CYP1A1 mRNA and protein expression are increased by benzo[a]pyrene. Wild ginseng moderately inhibits benzo[a]pyrene-induced CYP1A1 gene expression. Meanwhile, GSTA2, GSTA3, and GSTM2 gene expressions are also significantly increased by wild ginseng. As a result of metabolic regulations through the inhibition of metabolic activation enzymes, wild ginseng was found to protect against benzo[a]pyrene-induced hepatotoxicity.

19.3.2 Herb–Drug Interaction

Several cases of hepatotoxicity due to the use of herbals in hepatitis-related liver diseases have been reported. The effect of herb is primarily due to its inhibition of drug metabolic enzymes such as cytochrome P450.

19.4 Beverage

Green tea contains high concentrations of polyphenols, which are powerful antioxidants with anticarcinogenic properties. Green tea is assumed to possess a greater preventive potential than black tea due to its five times higher concentration of epigallocatechin-3-gallate. Molecular mechanisms underlying the chemoprevention effects exerted by green tea have been extensively investigated. Activation of Nrf2 is considered an important molecular target of chemoprotective agents.

Nevertheless, the balance between the inhibition of activation enzymes and the induction of detoxification enzymes is also an important consideration in determining the chemoprevention effects of green tea polyphenols. The effects of green tea polyphenols on metabolic enzymes are further described below.

19.4.1 Epigallocatechin-3-Gallate

Green tea polyphenols, especially epigallocatechin-3-gallate, were found to modulate metabolic detoxification enzymes such as glutathione S-transferase, glutathione peroxidase, glutamate cysteine ligase, and hemeoxygenase-1. These enzymes are involved in the detoxification of reactive chemical intermediates, including reactive oxygen species.

19.4.2 Polyphenols

Besides inducing detoxification enzymes, green tea polyphenols were also shown to exert cancer-protective activity by inhibiting the metabolic activation of carcinogens, particularly CYP450. Accordingly, tea polyphenols were reported to protect against tumors of the lung, gastrointestinal tract, and liver. An association was also found between green tea intake and the risk of esophageal cancer.

19.5 Alcohol

The roles of alcohol dehydrogenase (ALDH2) and CYP2E1 in promoting liver disease have been investigated. The accumulation of alcohol is a potentially pathologic condition that can progress to inflammation and carcinogenesis. During the liver disease process, CYP2E1 and CYP4A isozymes can be induced by alcohol. Activation of these CYP450 isozymes can produce reactive oxygen or nitrogen species, leading to oxidative modifications of DNA, proteins, and lipids.

19.5.1 Acetaldehyde

Among various potential mechanisms that could explain alcohol carcinogenicity is the metabolism of ethanol to acetaldehyde. Acetaldehyde is carcinogenic, which can lead to the formation of DNA adducts.

19.5.2 Curcumin Protective Effect

Studies were carried out on the hepatoprotective effects of a low dose of curcumin against liver damage induced by chronic alcohol. Curcumin was found to significantly reverse the alcohol–induced inhibition of alcohol and aldehyde dehydrogenase activities. Meanwhile, the activities of antioxidant enzymes and CYP4502E1 as well as the promotion of lipid peroxidation were also affected. Such findings indicate that low doses of curcumin may protect against liver damage caused by chronic alcohol intake.

19.6 Algae

Research on chlorophyll, unsaturated fatty acid, and polysaccharides isolated from algae has attracted interest due to their potential biological activities. Such algal extracts are briefly described below. Chlorophyll, unsaturated fatty acids, and lipopolysaccharides isolated from algae as inducers of detoxification enzymes are also included in Table 19.1.

19.6.1 Chlorophyll

Algae contain a powerful nutrient called chlorophyll, which is the key photosynthetic compound present in the plants. Methanolic extract of seaweed was shown to revert dioxin-induced toxicity through an alteration in the expression of antioxidants such as catalase, superoxide dismutase, glutathione, and glutathione peroxidase. Moreover, red and brown algae were found to modulate the expression of detoxification enzyme glutathione transferase against the substrate aryl halides.

19.6.2 Unsaturated Fatty Acid

Unsaturated fatty acid derivatives isolated as active components of green alga were shown to induce the expression of NAD(P)H:quinone oxidoreductase, heme oxygenase 1, and thioredoxin reductase. These findings led to propose that unsaturated fatty acid derivatives from green alga are able to activate the transcription factor that recognizes the antioxidant-response element.

19.6.3 Polysaccharide

Exploration of the chemoprevention activity of water-soluble polysaccharide extracts isolated from brown algae revealed that such polysaccharide extracts inhibit the activation enzyme CYP4501A as well as induce the detoxification enzyme glutathione-S-transferases.

On the other hand, studies of the effects of lipopolysaccharides produced by blue-green algae cyanobacteria on detoxification enzymes revealed a significant reduction of glutathione-S-transferase in embryos exposed to cyanobacteria. Further investigations are needed to explore the health benefits of algae.

19.7 Dietary Inducer–Drug Interactions

Enzyme induction or inhibition that increases or decreases the metabolism of drugs may affect the rate and the extent of medications. Such knowledge is essential to help minimize the potential adverse drug interactions. Understanding inducer–drug interaction helps research to explore the potential of utilizing metabolic enzyme inducers for chemoprevention.

19.7.1 Grapefruit

CYP450 enzymes are the major activation enzymes essential for the metabolism of many drugs. CYP450s can be inhibited or induced by drugs, resulting in significant interactions that can cause unanticipated adverse reactions. In the study of interaction between grapefruit juice and felodipine, grapefruit juice was found to reduce the metabolism of felodipine through selective regulation of CYP3A4 expression in the intestinal wall.

19.7.2 Sulforaphane

There are concerns that the antioxidative properties of sulforaphane may interfere with cytotoxic drugs such as doxorubicin. Doxorubicin is a cardiotoxic anticancerous drug that is used for a broad range of cancers. Sulforaphane was found to protect the heart from doxorubicin-induced toxicity and damage, leading to the enhancement of the efficacy of doxorubicin in the regression of tumor growth. Simultaneous administration with sulforaphane may reduce the dosage of doxorubicin.

19.7.3 Curcumin

Curcumin is another compound that has been studied for inducer–drug interaction. Due to its antioxidant and anti-inflammatory properties, curcumin has potential to be utilized for chemoprevention. Hence, curcumin has been studied for its effect on cardiomyocytes and cancer cells. Dose-dependent cardiotoxicity of doxorubicin could be reduced if supplemented with curcumin.

Moreover, the regulation of polyamine metabolism in human gastric and colon carcinoma cell lines in response to curcumin was also investigated to provide insight into the polyamine-related mechanisms and the potential of curcumin as the chemoprevention agent.

Bibliography

Amararathna M, Johnston MR, Rupasinghe HP (2016) Plant polyphenols as chemopreventive agents for lung cancer. Int J Mol Sci 17(8):1352

Angeloni C, Leoncini E, Malaguti M et al (2009) Modulation of phase II enzymes by sulforaphane: implications for its cardioprotective potential. J Agric Food Chem 57(12):5615–5622

Appelt LC, Reicks MM (1997) Soy feeding induces phase II enzymes in rat tissues. Nutr Cancer 28(3):270–275

Bailey DG, Malcolm J, Arnold O et al (1998) Grapefruit juice–drug interactions. Br J Clin Pharmacol 46(2):101–110

Balbo S, Philip J, Brooks PJ (2015) Implications of acetaldehyde-derived DNA adducts for understanding alcohol-related carcinogenesis. Adv Exp Med Biol 815:71–88

Bishayee A, Bhatia D, Thoppil RJ et al (2011) Pomegranate-mediated chemoprevention of experimental hepatocarcinogenesis involves Nrf2-regulated antioxidant mechanisms. Carcinogenesis 32(6):888–896

Bolling BW, Parkin KL (2008) Phenolic derivatives from soy flour ethanol extract are potent in vitro quinone reductase (QR) inducing agents. J Agric Food Chem 56(22):10473–10480

Bose C, Awasthi S, Sharma R et al (2018) Sulforaphane potentiates anticancer effects of doxorubicin and attenuates its cardiotoxicity in a breast cancer model. PLoS One 13(3):e0193918

Brooks PJ, Theruvathu JA (2005) DNA adducts from acetaldehyde: implications for alcohol-related carcinogenesis. Alcohol 35(3):187–193

Chen C-H (2012) Activation and detoxification enzymes: functions and implications. Springer Sciences, New York

Chen C-H (2020) Xenobiotic metabolic enzymes: bioactivation and antioxidant defense. Springer Nature, Cham

Chow HH, Hakim IA, Vining DR et al (2007) Modulation of human glutathione s-transferases by polyphenone intervention. Cancer Epidemiol Biomarkers Prev 16(8):1662–1666

Galijatovic A, Otake Y, Walle UK et al (2001) Induction of UDP-glucuronosyltransferase UGT1A1 by the flavonoid chrysin in Caco-2 cells—potential role in carcinogen bioinactivation. Pharm Res 18(3):374–379

Gamal-Eldeen AM, Ahmed EF, Abo-Zeid MA (2009) In vitro cancer chemopreventive properties of polysaccharide extract from the brown alga, Sargassum latifolium. Food Chem Toxicol 47(6):1378–1384

Gum SI, Jo SJ, Ahn SH et al (2007) The potent protective effect of wild ginseng (Panax ginseng C.A. Meyer) against benzo[alpha]pyrene-induced toxicity through metabolic regulation of CYP1A1 and GSTs. J Ethnopharmacol 112(3):568–576

Guo Z, Smith TJ, Wang E et al (1992) Effects of phenethyl isothiocyanate, a carcinogenesis inhibitor, on xenobiotic-metabolizing enzymes and nitrosamine metabolism in rats. Carcinogenesis 13(12):2205–2210

Herve C, de Franco PO, Groisillier A et al (2008) New members of the glutathione transferase family discovered in red and brown algae. Biochem J 412(3):535–544

Hodges RE, Minich DM (2015) Modulation of metabolic detoxification pathways using foods and food-derived components: a scientific review with clinical application. J Nutr Metab 2015: Article ID 760689

Ioannides C, Konsue N (2015) A principal mechanism for the cancer chemopreventive activity of phenethyl isothiocyanate is modulation of carcinogen metabolism. Drug Metab Rev 47(3): 356–373

Jain A, Rani V (2018) Assessment of herb-drug synergy to combat doxorubicin induced cardiotoxicity. Life Sci 205:97–106

Kallifatidis G, Labsch S, Rausch V et al (2011) Sulforaphane increases drug-mediated cytotoxicity toward cancer stem-like cells of pancreas and prostate. Mol Ther 19(1):188–195

Kensler TW (2001) Role of phase 2 enzyme induction in chemoprotection by dithiolethiones. Mutat Res 480-481:305–315

Kensler TW, Curphey TJ, Maxiutenko Y et al (2000) Chemoprotection by organosulfur inducers of phase 2 enzymes: dithiolethiones and dithiins. Drug Metabol Drug Interact 17:3–22

Keum YS (2011) Regulation of the Keap1/Nrf2 system by chemopreventive sulforaphane: implications of posttranslational modifications. Ann N Y Acad Sci 1229:184–189

Keum YS, Jeong WS, Kong AN (2005) Chemopreventive functions of isothiocyanates. Drug News Perspect 18(7):445–451

Lampe JW (2007) Diet, genetic polymorphisms, detoxification, and health risks. Altern Ther Health Med 13(2):S108–S111

Lampe JW, Chen C, Li S et al (2000) Modulation of human glutathione S-transferases by botanically defined vegetable diets. Cancer Epidemiol Biomarkers Prev 8:787–793

Lee H-I, McGregor RA, Choi M-S et al (2013) Low doses of curcumin protect alcohol-induced liver damage by modulation of the alcohol metabolic pathway, CYP2E1 and AMPK. Life Sci 93(18–19):693–699

Lee SE, Yang H, Son GW et al (2015) Eriodictyol protects endothelial cells against oxidative stress-induced cell death through modulating ERK/Nrf2/ARE-dependent heme oxygenase-1 expression. Int J Mol Sci 16(7):14526–14539

Leoncini E, Malaguti M, Angeloni C et al (2011) Cruciferous vegetable phytochemical sulforaphane affects phase II enzyme expression and activity in rat cardiomyocytes through modulation of Akt signaling pathway. J Food Sci 76:H175–H181

Liu D, Zhang L, Duan LX et al (2019) Potential of herb-drug/herb interactions between substrates and inhibitors of UGTs derived from herbal medicines. Pharmacol Res 150:104510

Maliakal P, Sankpal UT, Basha R et al (2011) Relevance of drug metabolizing enzyme activity modulation by tea polyphenols in the inhibition of esophageal tumorigenesis. Med Chem 7(5):480–487

Mastaloudis A, Wood SM (2012) Age-related changes in cellular protection, purification, and inflammation-related gene expression: role of dietary phytonutrients. Ann N Y Acad Sci 1259:112–120

Metere A, Giacomelli L (2017) Absorption, metabolism and protective role of fruits and vegetables polyphenols against gastric cancer. Eur Rev Med Pharmacol Sci 21(24):5850–5858

Moon YJ, Wang X, Morris ME (2006) Dietary flavonoids: effects on xenobiotic and carcinogen metabolism. Toxicol In Vitro 20(2):187–210

Morimitsu Y, Nakagawa Y, Hayashi K et al (2002) A sulforaphane analogue that potently activates the Nrf2-dependent detoxification pathway. J Biol Chem 277:3456–3463

Munday R, Munday CM (2001) Relative activities of organosulfur compounds derived from onions and garlic in increasing tissue activities of quinone reductase and glutathione transferase in rat tissues. Nutr Cancer 40(2):205–210

Munday R, Munday JS, Munday CM (2003) Comparative effects of mono-, di-, tri-, and tetrasulfides derived from plants of the Allium family: redox cycling in vitro and hemolytic activity and Phase 2 enzyme induction in vivo. Free Radic Biol Med 34(9):1200–1211

Munday R, Zhang Y, Munday CM et al (2006) Structure-activity relationships in the induction of Phase II enzymes by derivatives of 3H-1,2-dithiole-3-thione in rats. Chem Biol Interact 160(2):115–122

Murray-Stewart T, Dunworth M, Lui Y et al (2018) Curcumin mediates polyamine metabolism and sensitizes gastrointestinal cancer cells to antitumor polyamine-targeted therapies. PLoS One 13(8):e0202677

Na HK, Surh YJ (2008) Modulation of Nrf2-mediated antioxidant and detoxifying enzyme induction by the green tea polyphenol EGCG. Food Chem Toxicol 46(4):1271–1278

Nishinaka T, Ichijo Y, Ito M et al (2007) Curcumin activates human glutathione S-transferase P1 expression through antioxidant response. Toxicol Lett 170:238–247

Noratto GD, Chew BP, Atienza LM (2017) Red raspberry (Rubus idaeus L.) intake decreases oxidative stress in obese diabetic (db/db) mice. Food Chem 227:305–314

Osawa T, Kato Y (2005) Protective role of antioxidative food factors in oxidative stress caused by hyperglycemia. Ann N Y Acad Sci 1043:440–451

Parvez MK, Rishi V (2019) Herb-drug interactions and hepatotoxicity. Curr Drug Metab 20(4):275–282

Perez JL, Jayaprakasha GK, Cadena A et al (2010) In vivo induction of phase II detoxifying enzymes, glutathione transferase and quinone reductase by citrus triterpenoids. BMC Complement Altern Med 10:51

Rose P, Whiteman M, Moore PK et al (2005) Bioactive S-alk(en)yl cysteine sulfoxide metabolites in the genus Allium: the chemistry of potential therapeutic agents. Nat Prod Rep 22(3):351–368

Saw CL, Guo Y, Yang AY et al (2014) The berry constituents, quercetin, kaempferol, and pterostilbene synergistically attenuate reactive oxygen species: involvement of the Nrf2-ARE signaling pathway. Food Chem Toxicol 72:303–311

Shih PH, Yeh CT, Yen GC (2007) Anthocyanins induce the activation of phase II enzymes through the antioxidant response element pathway against oxidative stress-induced apoptosis. J Agric Food Chem 55:9427–9435

Singletary KW, Jung KJ, Giusti M (2007) Anthocyanin-rich grape extract blocks breast cell DNA damage. J Med Food 10(2):244–251

Song B-J, Akbar M, Jo I et al (2015) Translational implications of the alcohol-metabolizing enzymes, including cytochrome P450-2E1, in alcoholic and nonalcoholic liver disease. Adv Pharmacol 74:303–372

Srivastava A, Akoh CC, Fischer J et al (2007) Effect of anthocyanin fractions from selected cultivars of Georgia-grown blueberries on apoptosis and phase II enzymes. J Agric Food Chem 55(8):3180–3185

Syed DN, Chamcheu J-C, Adhami VM et al (2013) Pomegranate extracts and cancer prevention: molecular and cellular activities. Anti Cancer Agents Med Chem 13(8):1149–1161

Tan XL, Spivack SD (2009) Dietary chemoprevention strategies for induction of phase II xenobiotic-metabolizing enzymes in lung carcinogenesis: a review. Lung Cancer 65:129–137

Valerio LG Jr, Kepa JK, Pickwell GV et al (2001) Induction of human NAD(P)H:quinone oxidoreductase (NQO1) gene expression by the flavonol quercetin. Toxicol Lett 119(1):49–57

Wang R, Paul VJ, Luesch H (2013) Seaweed extracts and unsaturated fatty acid constituents from the green alga Ulva lactuca as activators of the cytoprotective Nrf2-ARE pathway. Free Radic Biol Med 57:141–153

Wu J-C, Lai C-S, Tsai M-L et al (2017) Chemopreventive effect of natural dietary compounds on xenobiotic-induced toxicity. J Food Drug Anal 25(1):176–186

Yang J, Liu RH (2009) Induction of phase II enzyme, quinone reductase, in murine hepatoma cells in vitro by grape extracts and selected phytochemicals. Food Chem 114:898–904

Yuan JM (2011) Green tea and prevention of esophageal and lung cancers. Mol Nutr Food Res 55(6):886–904

Zarfeshany A, Asgary S, Javanmard SH (2014) Potent health effects of pomegranate. Adv Biomed Res 3:100

Index

A

Acetaldehyde, 3, 17, 50, 108, 114, 135, 137, 143, 204, 205, 232, 238
Acetaminophen
  toxicity, 14, 99, 110
Acetyl-CoA (AcCoA), 17, 38, 65, 85, 86, 90
Acetylene, 71, 168, 169, 192
Acetyltransferase
  conjugation reaction, 81, 82
Activator, 45, 46, 153
Active
  transport, 21, 24–26, 30, 204
Acyltransferase
  conjugation reaction, 64
Aflatoxin B I
  toxicity, 12, 99, 145
Alcohol
  alcoholism, 135, 137, 143–144
  dehydrogenase, 17, 45–47, 50, 52–55, 71, 77, 96, 143, 204, 219, 238
  reaction, 53, 55
Aldehyde
  α,β unsaturated, 12
  dehydrogenase, 17, 49, 50, 52, 132, 137, 143, 219, 238
  oxidase, 46, 47, 49, 50, 53, 75, 76, 135
Algae, 101, 229, 230, 232, 239
Alkyl
  sulfides, 233
Amide, 35, 52, 55, 56, 63, 64, 67, 74, 85, 91, 137
Amine
  oxidase, 45–47, 49, 71, 75
  reaction, 56
Antioxidant
  activities, 175, 215, 216
  enzymes, 3, 41, 102, 103, 114, 119, 126, 238

Antioxidant response element (ARE), 5, 120, 128, 151, 153, 155, 156, 158, 159, 215, 229
Arsenic, 2, 15, 16, 64, 113, 115
ATP
  transporter, 25, 26, 28, 113
Azo
  dye, 2, 9, 11, 165
Azoreductase
  reaction, 51, 54

B

Benzene, 2, 12, 13, 15, 46, 115, 135, 143
Benzo[a]pyrene
  toxicity, 157
Beverage, 49, 182, 229, 230, 232, 237–238
Bifunctional
  inducer, 164, 165, 223–224
Bioactivation, 11, 48, 63, 75, 95, 96, 115, 123–125, 127, 132, 141, 151, 152, 154, 156, 230
Biomolecule, 121–123, 125–126, 203, 208, 210
Biphenyl, 2, 15, 16, 23, 46, 115

C

Cancer, 6, 10, 11, 13, 15, 16, 40, 100, 107–109, 111, 119, 121, 124, 126, 135–137, 139–145, 153, 157, 158, 182, 183, 203, 207–208, 210, 211, 219, 224, 225, 230, 233, 234, 236, 238, 240
Carbocation, 3, 100, 124, 146
Carbonium ion, 40, 95, 98, 100, 125, 136
Carboxylesterase, 45, 46, 51, 52, 55, 56, 71, 78, 98, 132, 137, 209
Carcinogen, 3, 10, 11, 18, 48, 74, 100, 109, 115, 132, 135, 136, 140–142, 146, 153, 158,

159, 164, 167, 175, 178, 179, 181, 182,
184, 191, 210, 211, 222, 224, 232, 234,
238
Carcinogenesis, 23, 64, 115, 116, 119, 120,
124–127, 132, 135, 138, 140, 141, 143,
145, 151, 156, 167, 178, 179, 182–184,
189, 194, 205, 210, 211, 222, 230, 231,
234, 235, 238
Carotenoid, 159, 184, 185, 232–235
Catalytic reactions
  phase I, 52, 98
  phase II, 81
Cellular components
  DNA adducts, 101
  lipid peroxidation, 101, 102, 120
  nucleic acid, 120
  protein adducts, 101, 120
Channel, 21, 24, 25, 113, 123, 224
Chemical
  carcinogenesis, 10, 48, 115, 135, 178, 182,
  189, 194, 222
Chemoprevention, 127–128, 134, 152,
157–159, 174, 179–182, 184,
210–211, 220, 223, 225, 234, 236,
237, 239, 240
Cigarette, 2, 3, 9, 13, 17–18, 109, 115, 164
Cinnamate, 165
Conformation, 167, 168, 189, 192, 215
Conjugation
  carboxylic acid, 64, 66, 85, 91
  C atom, 65, 90
  enzyme, 168
  N atom, 65, 89, 90
  O atom, 64, 88, 89
  OH group, 67
  S atom, 66, 90
Coumarin, 165
Curcumin, 157, 159, 169, 170, 182, 183, 198,
225, 232–235, 238, 240
CYP1A1, 74, 133, 134, 165, 180, 183, 184,
191, 221, 224, 237
CYP2A6, 48, 134–136
CYP2E1, 108, 109, 114, 135, 136, 164, 183,
184, 204, 221, 233, 238
Cytochrome P450 (CYP450)
  dealkylation, 71, 74
  dehydrogenation, 71, 73
  epoxidation, 71
  hydroxylation, 71, 72
  oxidation, 74
  reaction, 71, 74, 75
Cytoprotection, 155–157

D
Defense
  against oxidative stress, 119–121, 127, 128,
  151–159
  foreign compound, 2, 29, 40, 68, 206
Dehydrogenation
  alcohol, 73
  aldehyde, 73
Diallyl
  disulfide, 183, 184, 233
  sulfide, 159, 183, 184, 221, 233, 235
  trisulfide, 183, 184, 233
Diazepam, 2, 14, 15, 23
Dibenz[a,h]anthracene, 11, 18
Diesel exhaust, 2, 15–16, 178
Dietary
  algae, 239
  beverage, 237–238
  fruit, 159, 235–236
  grapefruit, 240
  inducer, 5, 48, 159, 167, 239–240
  vegetable, 159, 230–235
Diffusion, 17, 21, 23–26, 28, 30, 204
Dioxin, 2, 9, 15–17, 114, 126
1,2-dithiole-3-thione (D3T), 5, 127, 159, 163,
169, 177–179, 195, 196, 218, 220, 230,
232
DNA
  adduct, 74, 100, 101, 115, 121, 134,
  144–146, 157, 191, 209, 236, 238
  cell components, 109–111
  damage, 4, 109, 111, 121, 123, 126, 139,
  141, 143, 207
Drug
  metabolism interference, 115–116

E
Electrophile
  disease, 126
  drug, 127
  electrophilic stress, 124–127
  foreign compound, 62, 124
  metabolite, 95, 125, 190
Environmental
  chemical, 2, 5, 9, 15, 38, 114, 135, 140, 141,
  173, 221
  factor, 41, 102, 115
Enzyme
  beta-naphthoflavone, 181
  canthaxanthin, 169, 184, 185, 200
  catechol, 141, 142, 206

conformation, 167, 192
curcuminoid, 170, 195, 233
daidzein, 169, 181, 182, 198
ethoxyquin, 169, 200, 201
flavonoids, 5, 157, 159, 165, 169, 170, 177, 180–182, 195, 209, 220, 232–236
function, 142
genistein, 159, 169, 181, 182, 198, 219, 220, 234
indole-3-carbinol, 5, 108, 159, 169, 177, 180, 222, 224, 230, 232
inducer, 5, 48, 131, 156–157, 163, 164, 167–170, 177, 194–201, 215, 218, 220–223, 225, 232, 240
inducibility, 4, 131, 132, 161–170, 173–185, 215
inhibitor, 6, 112, 178, 179, 185, 205, 217, 221, 233
modulation, 5, 6, 157, 162, 163, 166–167, 175–177, 179–185, 216–220, 230–235
phenethyl isothiocyanate, 230, 231
polymorphisms, 1, 4, 41, 102, 131–146, 208, 210, 229–231
quercetin, 157, 159, 169, 170, 181, 196, 197, 232, 235
resveratrol, 159, 182, 198, 219
Epoxidation
epoxide, 36, 54, 72
epoxide hydrolase, 46, 52
reaction, 72
Excretion
foreign compound, 2, 21–31, 84, 95, 103, 161, 173, 174
Excretor, 31

F
Flavin-monooxygenase
reaction, 71, 72, 75, 98
Flavonoid
anthocyanin, 157, 181, 235, 236
beta-naphthoflavone, 181
4'-bromoflavone, 181, 223, 224
catechin, 169, 170, 181, 196, 197, 234
epigallocatechin, 169, 170, 181, 196, 197
isoliquiritigenin, 169, 181, 197, 223
leucocyanidine, 169, 170, 181
myricetin, 169, 170, 181, 196, 197
quercetin, 209, 234, 235
Foreign compound
radical, 62, 107
toxicity, 107, 133, 176

Free radical, 3, 40, 62, 95–98, 101–104, 107, 112, 114, 119–121, 124–126, 151, 153, 155, 156, 161, 167, 168, 204, 215, 229
Fruit
eriodyctiol, 235
polyphenol, 235, 236
pomegranate, 159, 232, 235, 236
procyanidin, 232, 235, 236
quercetin, 235
triterpene, 232, 235, 236
Functionalization
functional group, 2, 10, 27, 34, 36, 37, 45, 46, 52, 59–64, 71, 81–84, 95, 115, 124, 165, 168, 169, 189, 190, 193, 209, 223, 233

G
Genetic
polymorphism, 1, 4, 41, 47, 102, 131–143, 208, 210
variation, 132, 136, 137, 139, 141, 143, 231
Ginseng, 232, 237
Glucuronic acid (GA), 16, 33, 37, 39, 41, 60, 61, 64–66, 83, 88–91, 103, 140
Glutathione, 6, 11, 12, 15, 16, 33, 36, 37, 39–41, 54, 60–62, 65, 81–84, 90, 96, 99, 102, 103, 108, 110, 112, 115, 125, 127, 138, 139, 141, 145, 152, 156–158, 164, 165, 175, 177–179, 181, 182, 185, 191, 192, 204–208, 210, 211, 219, 220, 222–224, 230–234, 236, 237, 239
Glutathione S-transferase
conjugation reaction, 82, 102, 145
polymorphism, 138
Grapefruit
drug interaction, 224, 225

H
Health
benefit, 4, 6, 215–225
dietary, 6
effects, 5, 9, 16, 17, 59, 229, 230, 234
Hepatotoxicity
acetaminophen, 14, 98, 109, 206
aflatoxin, 180
alcohol, 164, 204
Herb
drug interaction, 232, 237
ginseng, 232, 237
rosemary, 222
tea, 159, 182, 237, 238

Heterocyclic
   amine, 2, 10, 140–142, 229
Household product
   benzene, 2, 13
Hydrolytic
   enzyme, 46, 51
Hydrophilic
   compound, 2, 21, 24, 26, 30, 33
Hydrophobic
   compound, 24, 208
Hydroquinone, 13, 39, 68, 87, 91, 142
Hydroxylation, 34–35, 46, 47, 71, 72, 145, 164, 175
4-hydroxynonenal, 12

I
Immune
   stimulation, 112, 114
   suppression, 112, 114
Indole-3-carbinol (I-3-C), 5, 108, 159, 169, 177, 180, 222, 224, 230, 232
Inducer
   activation, 164, 179, 216, 221
   bifunctional, 164, 165, 223–224
   detoxification, x
   dietary, 5, 167, 230, 232, 239–240
   monofunctional, 164, 165, 222–224
Inducibility, 4, 131, 132, 161–170, 173–185, 215
Inflammation, 3, 12, 15–17, 119, 120, 123–124, 126, 128, 139, 151–153, 156–158, 204, 205, 215, 229, 238
Intermediate
   formation, 95, 96, 98–100, 115
   reactive, 1, 3–6, 15, 23, 24, 40, 41, 47, 74, 81, 95–103, 107, 108, 110–112, 115, 116, 120–122, 124–126, 132, 136, 137, 142, 151, 152, 156, 166–168, 170, 173–175, 194, 203, 204, 206, 209, 210, 215–217, 219–221
Intrinsic
   toxicity, 161, 173
Ion
   transporter, 112, 113, 123
4-Ipomeanol
   toxicity, 40, 145–147
Isoflavones
   daidzein, 182
   genistein, 181, 182, 220
   soy, 159, 181, 182
Isoliquiritigenin, 169, 181, 197, 223

Isothiocyanate, 5, 62, 83, 127, 159, 165, 169, 177–179, 192, 195, 196, 211, 220, 222, 223, 230–232

K
Keap1-ARE pathway, 154, 155, 157, 215
Kidney, 3, 6, 11, 24, 27, 29, 30, 41, 48, 49, 60, 61, 64, 82, 85, 107, 110, 114, 123, 158, 182, 206–207

L
Lifestyle
   modification, 3, 102, 164
Lipid
   bilayer, 21, 24, 25, 111, 123, 203
   peroxidation, 3, 12, 100–102, 108, 110–112, 120, 121, 123, 125, 151, 152, 238
   solubility, 21, 23
Lipophile
   foreign compound, vii, 2, 21–23, 26–30, 36, 40, 46, 61, 71, 81, 95
   lipophilic, vii, 1, 4, 15, 21, 24, 27–30, 46, 60, 62, 115
Lipoxygenase, 45–47, 49, 50, 108

M
Menadione, 2, 14, 15, 23, 39, 98
Metabolic
   conversion, 1, 11, 14, 15, 21, 28, 29, 33–41, 95, 103, 108, 131, 151, 204
   intermediate, 3, 4, 6, 24, 27, 49, 59, 95, 99–102, 110, 122, 190, 206–207, 216
   metabolite, 1, 3, 4, 24, 27, 29, 33, 40, 59, 95, 102, 115, 127, 138, 191, 203, 204, 215, 221, 229
Methyltransferase
   conjugation, 63, 66, 67, 82, 88
   reaction, 63, 66, 67, 82, 86, 88–90
Michael
   donor, 193
   reaction acceptor, 168–169, 193
Mitochondria
   function, 112, 126
Modulation
   activation enzyme, 162, 163, 176, 216, 217
   detoxification enzyme, 163, 177, 218, 219
   interaction, 5, 191–192
   modulator, 5, 6, 189–200, 215, 216, 230

Molybdenum
  hydroxylase, 71, 76
  reaction, 72
Monofunctional
  cinnamate, 165
  coumarin, 165
  1,2-dithiol-3-thione, 165
  inducer, 164, 165, 222–224
  isoliquiritigenin, 223
  isothiocyanate, 165, 223
  phenol antioxidant, 165
  thiocarbamate, 165
Mycotoxin, 2, 9, 12, 98

N

N-acetyltransferase, 63, 82, 85–86, 99, 138, 141, 183
Neurodegeneration, 108, 119, 120, 124, 152, 153, 203, 208
Nicotine, 17, 34, 107, 135, 136
Nitrenium, 95, 97, 98, 100, 125
Nitroreductase, 45, 46, 51, 54, 55, 71, 72, 76
Nitrosamine, 2, 3, 9–11, 17, 109, 135, 136, 164, 178, 230
Non-conjugation
  catalytic reaction, 39, 68, 88–91
  enzyme, 36, 60, 67, 68, 81, 87, 99
  epoxide hydrolase, 39, 67–69, 87, 88, 91
Nrf2-ARE pathway, 5, 119, 120, 128, 151–159, 236
Nuclear transcription factor 2 (Nrf), 5, 119, 120, 128, 151–159, 203, 211, 215, 229, 232, 237
Nucleophile
  metabolite, 95
  nucleophilic, 48, 49, 61, 62, 82, 100, 103, 115, 168, 190, 191, 193, 210, 221
Nucleophilic metabolites, 61, 82, 190, 191

O

Olefin, 71, 168, 169, 192, 207
Oltipraz, 128, 159, 163, 165, 177–179, 196, 206, 211, 218, 224
Organosulfur
  alliin, 183
  compound, 5, 127, 177, 178, 183–184, 211, 221, 230, 232
Ortho hydroxyl, 170, 194
Overdose
  drug, 14, 110, 206

Oxidation
  compound, 47–49, 71, 72, 74, 75, 96, 131
Oxidative
  damage, 4, 109, 110, 121, 122, 126, 127, 204, 210, 222
  DNA, 4, 109, 111, 121, 123
  enzyme, 46–48, 50, 75
  lipid, 111
  protein, 4, 109, 110, 121, 122
  stress, 3, 4, 6, 14, 108, 109, 111, 112, 114, 115, 119–128, 138, 139, 142, 151–159, 164, 167, 203, 204, 206–208, 211, 215, 229, 234, 236

P

Passive
  diffusion, 21, 24, 25
Patulin, 12
Peroxidase
  reaction, 52, 53, 77, 103
Pharmaceutical, 1, 9, 13–15, 47, 51, 60, 137
Phase I
  activation, 2, 23, 27, 34–37, 45–56, 71, 81, 96, 98, 115, 124, 125, 131, 162, 164, 165, 176, 177, 180, 184, 190, 191, 209
  enzyme, 2, 4, 40, 41, 45–48, 52, 59, 71, 82, 95, 96, 98, 99, 108, 125, 127, 165, 194
  metabolism, 27, 33, 34, 47, 59, 61, 83, 96, 224
Phase II
  detoxification, 2, 23, 27, 36, 37, 40, 59–69, 81, 88, 103, 124, 128, 131, 138, 152, 158, 162, 165, 167, 174–176, 178, 180, 184, 191, 192, 205, 206, 210, 219, 223, 224
  enzymes, 2, 4, 41, 45, 52, 59, 60, 64, 68, 82, 96, 98, 99, 108, 127, 156, 163, 165–167, 174, 178, 183, 194–201, 223, 224
  metabolism, 4, 27, 33, 59–62, 81, 82, 84, 95, 96
Phase III
  metabolism, 28, 33, 60, 61
Phenobarbital, 162, 175, 185, 216, 219, 221
Phenol
  hydroxyl group, 170, 195
3-phosphoadenosine 5-phosphate, 84
3-phosphoadenosine 5-phosphosulfate, 39, 62, 67, 84, 89, 191
Phthalate, 2, 9, 13, 23
Polychlorinated
  biphenyl, 2, 15, 16, 23, 115

Polycyclic
  aromatic hydrocarbon, 2, 9, 11, 15, 17, 18, 48, 50, 54, 62, 75, 82, 125, 134, 139, 140, 143, 164, 182, 216, 219, 221, 229
Polymorphism
  enzyme, 41, 131–146, 230
  genetic, 1, 4, 41, 47, 102, 131–143, 208, 210, 229
Polyphenol
  carnosic acid, 169, 183
  carnosol, 183
  curcumin, 183
  ellagic acid, 169, 183
  gallic acid, 169, 183
  protocatechuic acid, 169, 170, 183, 198
  resveratrol, 169, 182, 183, 198
  tannic acid, 183
  turmeric, 183
Prostaglandine H synthase, 71, 98
Protein
  adduct, 100, 101, 121, 122, 152
  damage, 110
Pyridine, 10
Pyrimidine, 10

Q
Quinone
  reaction, 36, 39, 68, 81, 87, 91
  reductase, 36, 39, 67, 68, 81, 87, 91, 158, 164, 165, 170, 178, 181, 182, 192, 194, 210, 219, 220, 222–224, 230, 231, 233, 234, 236

R
Re-absorption
  kidney, 30
Reactive intermediate
  free radical, 3, 95, 102, 103, 112, 121, 125, 151, 161
  nitrogen species, 97, 108, 111, 119, 151, 155, 238
  phase I enzyme, 40, 99
Reactive oxygen species (ROS), 3, 6, 40, 41, 97, 101, 103, 108, 110, 111, 114, 119, 120, 122, 123, 125, 126, 136, 139, 140, 142, 153, 156, 158, 164, 167, 168, 175, 205–207, 215, 229, 234, 238
Reductive
  enzyme, 46, 51
  reaction, 47, 54, 55

Renal
  excretion, 29, 30, 205, 206
Resveratrol, 159, 169, 182, 183, 198, 219
Ribonucleotide
  reductase, 71, 77

S
S-adenosyl-L-homocysteine, 67, 87, 89
S-adenosylmethionine, 67, 87, 89, 141
Semiquinone, 39, 68, 87, 98
Site action, 23–24, 27
Smoke
  enzyme polymorphism, 143, 144
  smoker, 134, 137, 144
Solubility, 1, 2, 21–24, 27–29, 33, 34, 37, 38, 59, 60, 63, 71, 81, 103, 205, 209, 210
Solute carrier, 25, 204
Source of foreign compounds, 101, 123
Species difference
  aflatoxin, 145
  4-ipomeanol, 146, 147
  tamoxifen, 145, 146
Substrate-enzyme interaction, 5
Sulforaphane, 5, 127, 159, 169, 177–179, 195, 196, 211, 222–224, 230–232, 240
Sulfotransferase
  conjugation reaction, 62, 81, 82, 84, 88, 138, 146

T
Tamoxifen
  toxicity, 145, 146
Tea, 159, 182, 183, 237, 238
Terfenadine, 2, 14, 23
Terpenes
  astaxanthin, 184, 185
  ß-carotene, 185
  canthaxanthin, 184, 185
  lycopene, 184, 185
  terpenoid, 5, 127, 169, 177, 184–185
  zerumbone, 184, 185
Tetrodotoxin, 113, 123
Thiocarbamate, 165
Tobacco
  carcinogen, 109, 115, 140, 164, 175, 178
Toxicity
  cell components, 109–111
  cellular functions, 112–114
  intrinsic, 15, 107–108, 161, 173
  lifestyle induced, 108–109

Transporter
  active transport, 21, 24–26, 30, 204
  carrier, 204
  channel, 25, 113
  diffusion, 21, 24, 25, 30

U

Unsaturated aldehyde, 2, 9
Uridine-diphosphate-glucuronosyltransferase
  (UGT), 6, 60, 63, 82, 83, 100, 138, 140,
  142, 145, 146, 162, 176, 179, 181–185,
  218, 219
  conjugation reaction, 60, 66, 88, 89, 91,
  103, 156, 181, 190, 210

V

Vegetable
  allium, 230, 232, 233, 235
  cruciferous, 159, 178, 180, 211, 222, 224,
  230–232
  dietary, 230–233
  root, 230, 232, 233, 235
Vitamin
  vitamin C, 103
  vitamin E, 41, 104

W

Wine, 182

X

Xanthine
  oxidase, 14, 39, 45–47, 50, 52, 76, 108
Xenobiotic
  toxicity, 5, 131, 134, 139, 161–163,
  165–167, 173–177, 181, 216–220